Power pneumatics

Power pneumatics

Michael J Pinches & Brian J Callear

PRENTICE HALL

LONDON   NEW YORK   TORONTO   SYDNEY   TOKYO   SINGAPORE
MADRID   MEXICO CITY   MUNICH   PARIS

First published 1997 by
Prentice Hall Europe
Campus 400, Maylands Avenue
Hemel Hempstead
Hertfordshire, HP2 7EZ
A division of
Simon & Schuster International Group

Typeset in 9.5/12pt Garamond
by PPS, London Road, Amesbury, Wilts

Printed and bound in Great Britain by
Redwood Books, Trowbridge, Wiltshire

Library of Congress Cataloging-in-Publication Data

Callear, Brian J.
  Power pneumatics / Brian J. Callear & Michael J. Pinches.
      p.   cm.
  Includes index.
  ISBN 0–13–489790–0
  1. Pneumatic machinery.   2. Hydraulic machinery.   3. Fluid power
technology.   I. Pinches, Michael J., 1931–   . II. Title.
TJ950.C35   1997   96–18766
621.2—dc20   CIP

British Library Cataloguing in Publication Data

A catalogue record for this book is available from
the British Library

ISBN 0–13–489790–0

---

1   2   3   4   5   01   00   99   98   97

Contents

Preface

The authors have been involved in the education and industrial aspects of pneumatics for over 25 years. Currently Brian Callear is a lecturer and programme tutor in pneumatics at the North Notts Fluid Power Centre, senior examiner City & Guilds of London Institute in pneumatics, External Verifier for the City & Guilds of London Institute 2340 Fluid Power Competences scheme and a member of the British Fluid Power Association's Pneumatic Technical Steering Committee. Michael Pinches was, until his retirement, Manager of the Automation Advisory Centre at Sheffield Polytechnic, (now Sheffield Hallam University).

There are many excellent books dealing with the theoretical aspects of compressible fluids and several publications by the manufacturers of pneumatic equipment dealing with the construction and application of their products. This book is an attempt to bridge the gap between the two extremes and provide an introduction to the industrial application of power pneumatics.

The book covers basic pneumatic theory, types and modes of operation of pneumatic components, together with examples of their application. Worked examples and exercises are used throughout to illustrate the calculation, circuitry and component selection involved in the design of pneumatic systems.

This book will be useful to engineers involved in the design, specification, purchase and operation of pneumatic equipment. A chapter on maintenance gives some guidance on the installation and operation of systems with a view to increasing reliability and reducing downtime.

The contents of the book cover the sylabii of the 'B Tec' and 'C&G I' pneumatic modules and most first-year degree courses. The subject-matter is covered in both a qualitative and quantitative manner and should prove useful as a guide and reference book for all lecturers and students.

Acknowledgements

The authors particularly wish to thank Lynda Callear for her patience and extreme hard work in translating, preparing and correcting the manuscript, Rebecca Callear for her sufferance and tireless supplies of tea and to Audrey Pinches for her encouragement and support while the book was being prepared.

The various manufacturers who have provided or permitted us to copy diagrams and tables from their literature are gratefully acknowledged.

Atlas Copco
Robert Bosch
Norgren
Rexroth
SMC
Compair Maxam

Introduction

The application of compressed air (pneumatics) dates back to the Bronze Age or earlier, when bellows made out of animal skin were used to create a draught for a fire. The word 'pneumatics' is derived from the Greek word *pneumos* meaning 'breath'. It was not, however, until the Industrial Revolution that any extensive use was made of compressed air, with some of the early applications being in mines for ventilation and pumping water. Pneumatically operated percussive rock drills were used in the construction of the Mount Censis Tunnel in the mid-nineteenth century.

The earliest attempts to transmit compressed air over a distance used clay pipes. Unfortunately the tests were a complete failure as the air leaked through the pipe joints. A system using pipework laid in the Paris sewers conveyed compressed air throughout the city from a central compressor station, where a 2000 horsepower installation supplied air at 85 lb/in^2 (6 bar). This system was initially very successful and the compressor capacity was considerably increased over the next three years. Its eventual demise resulted from the introduction of a more economical electrical supply network. However, the ready availability of electricty led to the development of compressed air systems powered by small local compressor plants. These units adopted the same operating pressure as the centralised supply and to this day a pressure of 7 bar (100 lb/in^2) is the value at which most modern industrial systems operate.

It was not until the early twentieth century that there was any real development in the industrial application of pneumatics. Mechanisation and automation considerably increased the usage of compressed air, and nowadays almost all industrial units have a compressed air system.

CHAPTER 1

Pneumatic principles

1.1 Units

There is still some confusion regarding the meanings of the term *mass* and *weight*. Several systems of units have been used to try to solve the problem, notably, in the imperial system: the slug and the pound, the pound and the poundal, the pound mass and the pound force. In the SI system the unit of mass is the *kilogram* and the unit of force is the *newton*, but before using these units an understanding of their meaning is essential. Every physical substance comprises molecules, each of which possesses mass. As any particular body will have a constant number of molecules, provided the body is not physically altered, mass is therefore a property of that body and is constant. The weight of the body is, however, variable depending upon its mass and the acceleration to which it is subjected. In an area of zero acceleration the body will have no weight. Astronauts are subjected to weightlessness when outside the gravitational field of the earth or any other planet.

Any body that is subjected to an acceleration will experience a force F in the direction of that acceleration. The value of the force F due to the acceleration is proportional to the product of the mass m of the body and the acceleration a to which the body is subjected; or

$$\text{Force} \propto \text{Mass} \times \text{Acceleration}$$

$$F = k \times m \times a$$

where k is a constant. By a careful selection of units the value of k will be unity, and

$$F = m \times a$$

where F is in newtons (N), m is in kilograms (kg) and a is in metres/second/second (m/s^2).

The weight of a body on earth is the force exerted on that body by the gravitational pull of the earth. The standard acceleration due to the earth's gravity is defined as 9.80665 m/s^2. This is an arbitrary value; the exact value of gravity varies by about 0.5 per cent over the earth's surface. The value generally used in engineering calculations is 9.81 m/s^2.

Example 1.1

A body has a mass of 1 kg. What is its weight
(a) on earth in a gravitational acceleration of 9.81 m/s^2?

(b) on a smaller planet 'Pneumo' whose gravitational acceleration is 6 m/s²?

Solution

Weight on earth = Mass × Acceleration

$$= 1 \text{ kg} \times 9.81 \text{ m/s}^2$$

$$= 9.81 \text{ kg m/s}^2$$

$$= 9.81 \text{ newtons (N)}$$

Note: 1 N = 1 kg m/s²

Weight on planet 'Pneumo' = Mass × Acceleration

$$= 1 \text{ kg} \times 6 \text{ m/s}^2$$

$$= 6 \text{ kg m/s}^2$$
$$= 6 \text{ N}$$

There is a reduction in weight when in a lower gravitational system. When astronauts 'walked' on the moon, they could jump much further than was possible on earth owing to their reduced weight.

In the past, force was usually expressed in terms of the gravitational pull on a known mass, loosely referred to as weight. This gave rise to the terms *kilogram weight* and *pound weight*. As gravity varies over the earth's surface, the value of these terms vary. Using the international standard gravity value, the kilogram force can be exactly defined, i.e. 1 kg = 9.80665 N.

The term *g force* is sometimes used; it is an abbreviation for gravitational force. The standard gravitational acceleration is referred to as an acceleration of 1*g*. Thus an acceleration of 2*g* would mean 2 × 9.80665 m/s². When astronauts 'blast off' from earth they are subjected to *g* forces in excess of 3 – i.e. three times the standard gravitational force N so their body weight will increase three-fold.

Example 1.2

A mass of 1 kg is suspended on a spring balance in a lift cage. The maximum acceleration of the lift both ascending and descending is 3 m/s². If the spring balance is calibrated in kg what will be its reading at maximum acceleration when
(a) ascending?
(b) descending?

Solution

Ascending (see Fig. 1.1)

With the cage ascending the spring balance will be accelerated upwards at a maximum rate of 3 m/s². To the spring balance it will appear that the mass is being

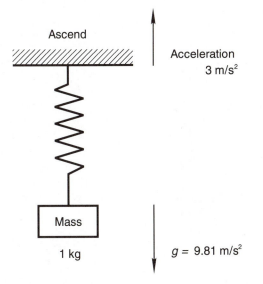

FIGURE 1.1 Acceleration of the cage upwards.

pulled downwards with an acceleration of 3 m/s^2 in addition to the gravitational acceleration.

Total acceleration is 9.81 + 3 m/s^2. Weight will be

$$1 \times (9.81 + 3)\ N = 12.81\ N$$

As the spring balance is calibrated in kg it will read

$$\frac{12.81\ N}{9.81\ m/s^2} = 1.306\ kg$$

Descending (see Fig. 1.2)

Effective acceleration on weight = (9.81–3) m/s^2. This time the reading on the spring balance will be

$$\frac{9.81 - 3}{9.81} = 0.694\ kg$$

1.1.1 Systems of units

Among the many systems of units currently in use the most widely used are:

1. The SI system, which uses the metre, kilogram and second as its base.
2. The imperial system, which uses the foot, pound and second.
3. The metric system, which is based on the metre, kilogram and second.

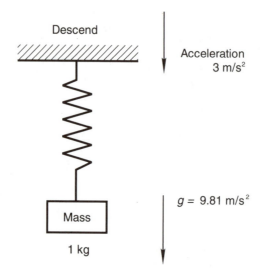

FIGURE 1.2 Acceleration of the cage downwards.

1.1.2 Pressure

This is defined as force per unit area.

$$\text{Pressure} = \text{Force (newtons)/Area (square metres)}$$

The SI unit, N/m^2, is called a Pascal (Pa), therefore $1\ N/m^2 = 1$ Pa. This is a very small unit of pressure and in pneumatics the bar is more commonly used, where $1\ \text{bar} = 10^5$ N/m^2.

    The megapascal or mPa is also being increasingly used for higher pressure systems where

$$1\ \text{mPa} = 10\ \text{bar} = 10^6\ N/m^2$$

The imperial system uses lb/in^2 and ton/in^2 to measure pressure, while in the metric system kg/cm^2 are the units used. The relationship between these units is shown in Table 1.1.

Atmospheric pressure
This is the pressure on the surface of the earth caused by the weight of the air in the atmosphere. Atmospheric pressure varies from place to place and with time. For most pneumatic calculations atmospheric pressure may be considered constant and equal to 1 bar or $10^5\ N/m^2$. When pressure is measured above atmospheric pressure it is referred to as gauge pressure.

TABLE 1.1 Currently used systems of units

Quantity	Symbol	SI System		Imperial System		Metric System	
Length	l	metre 1 m = 39.37 in 1 m = 3.281 ft	m	inch 1 in = 0.0254 m foot 1 ft = 0.3048 m	in ft	centimetre 1 cm = 10^{-2} m millimetre 1 mm = 10^{-3} m	cm mm
		micron 1 μm = 10^{-6} m	μm				
Area	A	square metre 1 m^2 = 1550 in^2	m^2	square inch 1 in^2 = 0.645 × 10^{-3} m^2 1 in^2 = 6.45 cm^2	in^2	square centimetre 1 cm^2 = 10^{-4} m^2	cm^2
Volume	V	cubic metre 1 m^3 = 220 gal 1 m^3 = 10^3 litres	m^3	cubic inch 1 in^3 = 16.39 × 10^{-6} m^3 gallon 1 gal = 277.4 in^3 = 0,00454 m^3 cubic foot 1 ft^3 = 6.24 gal	in^3 gal ft^3	Cubic centimetre 1 cm^3 = 10^{-6} m^3 Litre 1 l = 10^{-3} m^3	cm^3 l
Time	t	second	s	minute	min	minute	min
Volumetric flow rate	q	cubic metres per second 1 m^3/s = 13.2 × 10^3 gal/min	m^3/s	cubic inches per minute gallons per minute	in^3/min gal/min	litres per minute	l/min

TABLE 1.1 (*Continued*)

Quantity	Symbol	SI System		Imperial System		Metric System	
Velocity	v	metres per second	m/s	feet per second	ft/s	metres per minute	m/min
Acceleration	a	metres per second squared	m/s²	feet per second squared	ft/s²	metres per second squared	m/s²
Mass	m	kilogram 1 kg = 2.2 lb	kg	pound mass 1 lb = 0.4536 kg	lb	kg s²/m = 9.807 kg	kg s²/m
Force or weight	F.P.	newton	N	pound force 1 lbf = 4.45 N	lbf	kilogram force 1 kp = 1 kgf = 9.81 N	kp kgf
Torque	M T	newton metre	N m	foot pound force 1 ft lbf = 1.356 N m	ft lbf	kilogram force metre 1 kp m = 1 kgf m = 9.81 Nm	kp m kgf m
Pressure	P	newton per square metre 1 bar = 10⁵ N/m² 1 Pa (Pascal) = 1 N/m²	N/m²	pound force per square inch 1 lbf/in² = 6897 N/m²	lbf/in²	kilogram force per square centimetre 1 kgf/cm² = 9.81 × 10⁴ N/m²	kgf/cm² kp/cm²
Work	A W	joule 1 J = 1 Nm	J	foot pound force 1 ft lbf = 1.356 J	ft lbf	kilogram force metre 1 kgf m = 9.81 J	kgf m kp m
Power	P N	watts 1 W = 1 N m/s	W	foot pound force per second 1 ft lbf/s = 1.356 W horse power 1 hp = 745.7 W	ft lbf/s hp	metric horse power 1 PS = 1 ch = 75 kp m/s = 735.5 W	PS ch

Absolute pressure

Should the pressure be measured above absolute vacuum it is known as absolute pressure, i.e.

$$\text{Absolute pressure} = \text{Gauge pressure}$$

$$+ \text{Atmospheric pressure}$$

The vast majority of pressure gauges are calibrated with atmospheric pressure as the zero point. In this system it is possible to have negative pressures up to minus 1 bar, indicating vacuum conditions.

Note that all calculations involving the gas laws require values of pressure and temperature to be in absolute units. In all other calculations gauge pressures are used.

Example 1.3

A pneumatic cylinder with a bore of 100 mm is to clamp a component with a static force of 3000 N (see Fig. 1.3). Determine the required air pressure.

Solution

$$\text{Area of cylinder} = \pi \times \frac{D^2}{4}$$

$$= \pi \times \frac{100^2}{4} \text{ mm}^2$$

$$= \pi \times \frac{0.1^2}{4} \text{ m}^2$$

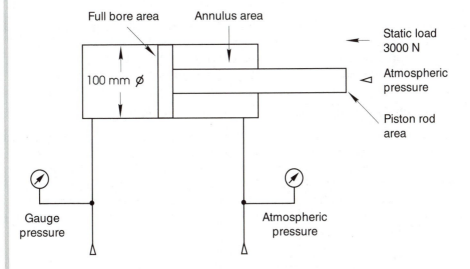

FIGURE 1.3 Arrangement of the clamp cylinder.

$$= 7.85 \times 10^{-3} \text{ m}^2$$

$$(\text{Note } 100 \text{ mm} = 0.1 \text{ m})$$

$$\text{System pressure} = \frac{\text{Force}}{\text{Area}}$$

$$= \frac{3000 \text{ N}}{7.85 \times 10^{-3} \text{ m}^2}$$

$$= 382 \times 10^3 \text{ N/m}^2$$

$$= 3.82 \times 10^5 \text{ N/m}^2$$

$$= 3.82 \text{ bar, where } 1 \text{ bar} = 10^5 \text{ N/m}^2$$

The system pressure acts on one side of the piston and is opposed by atmospheric pressure acting on the other side. The system pressure has therefore to overcome the atmospheric pressure and exert a thrust of 3000 N. The pressure registered on a gauge is the pressure above atmospheric.

Taking atmospheric pressure as 1 bar absolute, then

$$\text{System pressure} = 3.82 \text{ bar gauge}$$

$$= 3.82 \text{ bar} + 1 \text{ bar absolute}$$

$$= 4.82 \text{ bar absolute}$$

Note: Unless a pressure is stated or referred to as absolute, it is usual to assume it is a gauge pressure.

1.1.3 Pascal's laws

Pascal's laws relate to the pressure in a fluid – covering both the liquid and gaseous states.

1. The pressure will be the same throughout an enclosed volume of fluid which is at rest, provided the effect of the weight of the fluid is neglected.
2. The static pressure acts equally in all directions at the same time.
3. The static pressure always acts at right angles to any surface in contact with the fluid.

1.1.4 Quantity flowing

As air is compressible its volume varies with pressure and temperature; therefore, when stating a flow rate, the pressure and temperature must also be given. To standardise it is usual to refer to the quantity of air at atmospheric pressure (taken as 1 bar abs) and at a temperature of 20 °C. This is known as 'free air'. The output of a compressor may be given as so many units of 'free air' delivered (f.a.d). Volumetric measurement in the SI system is stated in litres (l) or cubic metres (m^3). In the imperial system the unit of volume is the cubic foot (ft^3).

A flow rate of 1 m^3/min (f.a.d) = 1000 l/min (f.a.d)

$$= 1000 \text{ dm}^3/\text{min (f.a.d)}$$

$$= 16.7 \text{ l/s (f.a.d)}$$

$$= 16.7 \text{ dm}^3/\text{s (f.a.d)}$$

$$= 35.26 \text{ ft}^3/\text{min (f.a.d)}$$

In pneumatic systems the cubic decimetre of free air per second is generally recognised as the standard unit for the measurement of air flow.

Note: 1 cubic decimetre $= 1 \text{ dm}^3 = 1$ litre

1.2 Properties of air

Air is the most plentiful material on earth and one of the most complex mixtures. For the purposes of pneumatics it may be considered to be a mixture of nitrogen and oxygen (78 per cent and 21 per cent by volume) with impurities of water vapour and dirt. It may be considered as a perfect gas for purposes of calculation over the pressure and temperature limits encountered in normal pneumatic circuits.

Physical constants of air:

Molecular weight $= 28.96 \text{ kg/kmol}$

Density at $15\,^{\circ}\text{C}$ and 1 bar $= 1.21 \text{ kg/m}^3$

Boiling point at 1 bar $= -191$ to $-194\,^{\circ}\text{C}$

Freezing point at 1 bar $= -212$ to $-216\,^{\circ}\text{C}$

Gas constant $= 286.9 \text{ J/kg K}$

Generally, atmospheric air is neither perfectly dry nor fully saturated with water vapour but at an intermediate condition. The ratio of the water present in the atmosphere to the water content in saturated air at the same temperature is known as the *relative humidity*. If wet air is cooled a temperature will be reached when moisture condenses out of the air; this temperature is dependent upon the relative humidity and is known as the *dew point*. The higher the air temperature, the greater the water content of saturated air. The water content of saturated air can be calculated by applying Dalton's Law of Partial Pressures. This states that the total pressure exerted by a mixture of gases and vapour is equal to the sum of the pressure of each gas and vapour taken separately, if the quantity of gas and vapour occupied the same volume as that of the mixture at the same temperature. Thus, in a mixture of air and water vapour:

Pressure of a mixture = Partial pressure of air

+ Partial pressure of water vapour.

Table 1.2 shows the mass of water in kilograms per 100 m^3 of free air at the temperatures and pressures shown.

TABLE 1.2 Mass of water (kg) per 100 m³ of free saturated air.

Temp. (°C)	Pressure (bar (gauge))				
	0	2	4	6	8
0	0.48	0.17	0.10	0.07	0.05
20	1.73	0.576	0.346	0.247	0.192
40	5.10	1.70	1.02	0.728	0.567
60	12.95	4.32	2.59	1.85	1.44
80	29.04	9.68	5.81	4.15	3.23
100	56.00	19.33	11.76	8.40	6.53
120	0	36.73	22.04	15.74	12.24
140	0	0	37.74	29.96	20.97

Example 1.4

A compressor delivers 500 m³ of free air per hour at a pressure of 7 bar gauge and a temperature of 40 °C. The atmospheric air at the compressor intake has a relative humidity of 80 per cent and a temperature of 20 °C. Determine the amount of water that has to be extracted from the compressor plant per hour.

Solution

At 20 °C and 0 bar (gauge), 100 m³ of free saturated air contains 1.73 kg of water. At 80 per cent humidity the water content is 1.384 kg per 100 m³ f.a.d. Air at the compressor plant outlet will be saturated. Water content per 100 m³ f.a.d at 7 bar gauge and 40 °C by interpolation is

$$\frac{0.728 + 0.567}{2} = 0.647 \text{ kg}$$

Thus water content of air entering the compressor plant per hour is

$$5 \times 1.84 \text{ kg}$$

Water content of air leaving the compressor plant per hour is

$$5 \times 0.647 \text{ kg}$$

Thus, the amount of water that has to be extracted from the compressor plant per hour is

$$5 \times (1.384 - 0.647) \text{ kg} = 3.68 \text{ kg (or 3.68 litres)}$$

1.2.1 The gas laws

In pneumatic systems a gas, normally air, is used to store, transport and give out energy. A compressor increases the pressure of the air which is then fed to some form of actuator and made to do work. In order to fully understand pneumatics a basic knowledge of the compression and expansion of a gas is required.

Strictly speaking the gas laws should only be applied to a perfect gas. Although air is a complex mixture, using the gas laws will give a sufficiently accurate answer in pneumatic applications. In all calculations involving gases, absolute values of pressure and temperature must be used.

Temperature measurement

The temperature of a substance is an indication of its hotness or coldness and is measured by some type of thermometer. Several scales have been used in the measurement of temperature. The one in current use is the Celsius (Centigrade) scale. This scale has a zero value at which pure water freezes at a pressure of 760 mm of mercury.

The absolute zero of temperature is the value at which all molecular movement in a perfect gas ceases, and it has a value of $-273\,°C$. To convert from $°C$ to $°C$ absolute add 273, thus

$$100\,°C = 100 + 273$$

$$= 373\,°C \text{ absolute}$$

Absolute temperature is given in $°C$, although absolute is sometimes referred to as K or Kelvin. Thus

$$373\,°C \text{ absolute} = 373 \text{ K}$$

The absolute zero of temperature is 0 K.

The Fahrenheit scale of temperature is still used to a minor extent. On this scale the temperature at which pure water freezes at a pressure of 760 mm mercury is $32\,°$ Fahrenheit or $32\,°F$.

To change from $°F$ to $°C$ use the formula

$$°C = (°F - 32) \times 5/9$$

For example,

$$80\,°F = (80 - 32) \times 5/9$$

$$= 26.7\,°C$$

Boyle's law

This states that in a perfect gas in which the mass and temperature remain constant, the volume V varies inversely as the absolute pressure P, or

$$PV = \text{constant}$$

Charles' law

This states that in a perfect gas in which the mass and the pressure remain constant, the volume V varies directly as the absolute temperature T:

$$\frac{V}{T} = \text{constant}$$

1.2.2 Characteristic gas equation

Boyle's and Charles' laws are combined into the characteristic equation for a perfect gas, which states

$$PV = mRT$$

where m = the mass of the gas

and R = the characteristic gas constant for the particular gas.

This equation may be rewritten

$$\frac{PV}{T} = mR$$

For a given mass of gas

$$mR = \text{constant}$$

Thus, for a given mass of gas, if there is a change of state from conditions noted by suffix 1 to those shown by suffix 2, then

$$\frac{P_1 V_1}{T_1} = \frac{P_2 V_2}{T_2}$$

1.3 Expansion and compression of gases

When the volume of a given mass of gas changes, that gas is being either compressed or expanded. The change in volume is accompanied by a change in pressure or temperature, or both. The way in which the pressure, temperature and volume change is dependent upon the flow of energy into or out of the gas. Provided there is no change in the mass of the gas, it will obey the characteristic gas equation if it is a perfect gas.

Note: A perfect gas is clean and dry.

1.3.1 Types of expansion and compression

A gas can be compressed or expanded in various controlled ways. These are:

1. *Isothermal*, at constant temperature the volume change for a given mass of gas follows the law

$$PV = \text{constant}$$

or

$$P_1 V_1 = P_2 V_2$$

where suffix 1 denotes initial conditions and suffix 2 denotes final conditions.

2. *Isentropic or adiabatic*, where there is no flow of heat energy either into or out of the gas during the expansion or compression. The gas law governing this process is:

$$PV^\gamma = \text{constant}$$

where γ is the ratio of the specific heats of the gas and is known as the adiabatic index:

$$\gamma = \frac{C_p}{C_v}$$

where C_p is the specific heat at a constant pressure and C_v is the specific heat at constant volume.

For air

$$C_p = 1.005 \text{ kJ/kg K and } C_v = 0.718 \text{ kJ/kg K}$$

$$C_p = 0.24 \qquad \text{and } C_v = 0.17$$

The specific heat ratio being 1.4 for air.

3. *Polytropic*, which is the most common type of volume change, lying between isothermal and isentropic. It is governed by the law:

$$PV^n = \text{constant}$$

where n is the polytropic index lying between 1 and 1.4 for all types of volume change. The universal gas law

$$PV = mRT$$

also applies.

For a change to be isothermal it must occur very slowly, so that the temperature of the gas remains unchanged. For a change to be adiabatic it must occur very rapidly so that there can be no flow of energy into or out of the system. In pneumatics most expansions and compressions take place somewhere between isothermal and adiabatic, i.e. polytropic.

PV diagrams

Consider a piston being driven forward at a constant pressure P, with compressed air being continually fed into the cylinder (Fig. 1.4).

Area of piston $\qquad = A$

Force exerted by the piston $= F$

where

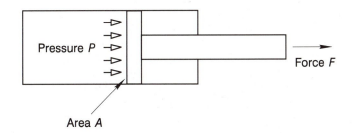

FIGURE 1.4 Piston being extended by a continuous supply of compressed air.

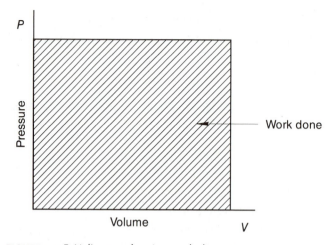

FIGURE 1.5 *P–V* diagram showing work done at constant pressure.

$$F = P \times A$$

If the piston moves a distance L

$$\text{Work done} = \text{Force} \times \text{Distance moved}$$

$$= P \times A \times L$$

but

$$A \times L = \text{Volume of air entering the cylinder}$$

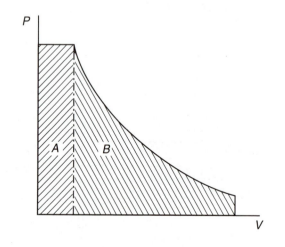

FIGURE 1.6 *P–V* diagram showing total work done at constant pressure and by expanding air.

So

$$\text{Work done} = P \times V$$

This can be shown on a *P–V* diagram (Fig. 1.5).

$$\text{The area under the curves} = P \times V$$

$$= \text{Work done}$$

In general the area under a *P–V* diagram represents the work done either by the gas, or on the gas, no matter what law governs the process.

In the *P–V* diagram shown in Fig. 1.6 the total work done is represented by the areas *A* + *B*.

1.4 **Air compression**

Consider the single-acting piston-type compressor shown in Fig. 1.7.

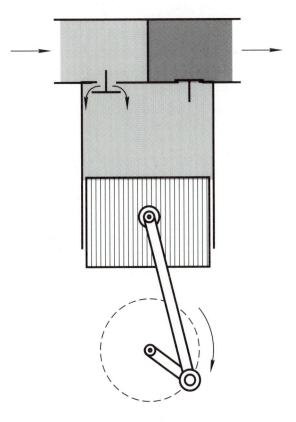

FIGURE 1.7 General arrangement of a single-acting single-cylinder compressor.

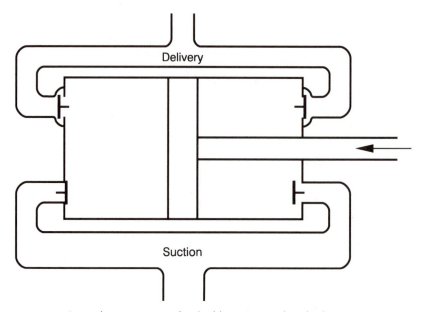

FIGURE 1.8 General arrangement of a double-acting single-cylinder compressor.

$$\text{Compressor swept volume for suction stroke} = \frac{\pi D^2 L}{4}$$

As the compressor is single acting there is one suction stroke and one delivery stroke per cycle. In a double-acting piston-type compressor, as shown diagrammatically in Fig. 1.8, there will be two suction and two delivery strokes per cycle.

In order to compare the performance of compressors under a range of different conditions, the air delivered is specified in terms of free air. The volumetric efficiency of the compressor, η_{vol}, is given by

$$\eta_{vol} = \frac{\text{Weight of free air delivered per minute}}{\text{Weight of air equivalent to swept volume at inlet conditions}}$$

$$= \frac{\text{Volume of free air delivered per minute}}{\text{Swept volume of compressor per minute}}$$

Volumetric efficiency depends upon the following factors:

1. The weight of air entering the compressor (which is affected by any restrictions in the inlet pipework), air temperature and pressure at the inlet. The higher the inlet temperature, the lower the weight of air entering the compressor per cycle.
2. The clearance volume at the ends of the cylinder which allows for the operation of the inlet and discharge valves (in the case of reciprocating compressors).
3. The internal leakage of the compressor.

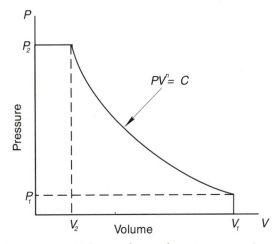

FIGURE 1.9 *P–V* diagram for a polytropic compression.

Compressor speed will also affect the volumetric efficiency. The higher the speed the less time is available for air to flow into the cylinder during the suction stroke, which results in a reduction in the weight of air per stroke entering the cylinder.

1.4.1 Work done during compression

During compression the theoretical work done depends upon the type of compression taking place – i.e. polytropic, isentropic or isothermal. Consider the pressure to volume or *P–V* diagram for polytropic compression (Fig. 1.9).

Air is taken into the compressor at a pressure P_1 to fill the swept volume V_1. It is compressed polytropically to the delivery pressure P_2 and the volume reduces to V_2. Work done during the cycle is represented by the area below the curve.

$$\text{Work done} = \int_{V_2}^{V_1} P \, dV + P_2 V_2 - P_1 V_1$$

Now

$$PV^n = C \quad \text{or} \quad P = CV^{-n}$$

Thus

$$\text{Work done} = \int_{V_2}^{V_1} CV^{-n} \, dV + P_2 V_2 - P_1 V_1$$

$$= \left[\frac{CV^{1-n}}{1-n} \right]_{V_2}^{V_1} + P_2 V_2 - P_1 V_1$$

$$= \frac{CV_1^{1-n} - CV_2^{1-n}}{1-n} + P_2 V_2 - P_1 V_1$$

$$= \frac{(P_2 V_2 - P_1 V_1) + (n-1)(P_2 V_2 - P_1 V_1)}{n-1}$$

$$= \frac{n}{n-1} \ (P_2 V_2 - P_1 V_1)$$

Example 1.5

A compressor delivers 3 m³ of free air per minute at a pressure of 7 bar gauge. Assuming that the compression follows the law $PV^{1.3} = $ constant, determine the theoretical work done.

Solution

$$\text{Work done} = \frac{n}{n-1} \ (P_2 V_2 - P_1 V_1)$$

where $n = 1.3$

P_1 = atmospheric pressure = 1 bar abs

$V_1 = 3$ m³/min

$P_2 = 7$ bar gauge = 8 bar abs

$$(P_1 V_1)^{1.3} = (P_2 V_2)^{1.3}$$

$$V_2 = V_1 \times \left(\frac{P_1}{P_2}\right)^{1/1.3}$$

$$= 3 \times \left(\frac{1}{8}\right)^{1/1.3}$$

$$= 0.61 \text{ m}^3/\text{min}$$

$$\text{Work done} = \frac{1.3}{0.3} \ (8 \times 0.61 - 1 \times 3) \times 10^5 \text{ N/m}^2 \times \text{m}^3/\text{min}$$

$$= 8.15 \times 10^5 \text{ N m/min}$$

$$= 13.58 \times 10^3 \text{ N m/s}$$

But 1 N m/s = 1 watt; therefore,

$$\text{Work done} = 13.58 \text{ kW}$$

The value of the polytropic index can vary (dependent upon the conditions of compression) between a minimum value of 1 and a maximum value of approximately 1.4. When the index has a value of 1 the compression is isothermal (constant temperature), i.e.

$$PV = \text{constant}$$

When the index has a value of 1.4 the compression is said to be isentropic, meaning that there is no change in the total heat energy of the gas under compression. A full explanation of the types of compression and expansion can be found in any basic thermodynamics textbook. The *P–V* diagram for isothermal compression is shown in Fig. 1.10.

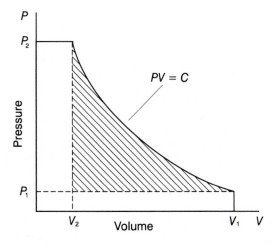

FIGURE 1.10 *P–V* diagram for an isothermal compression.

The work done per stroke is represented by the shaded area and is given by:

$$\text{Work done} = \int_{V_2}^{V_1} P \, dV + P_2 V_2 - P_1 V_1$$

Note $P_1 V_1 = P_2 V_2 = C$ and $P = C/V$. Therefore,

$$\int_{V_2}^{V_1} P \, dV = C \int_{V_2}^{V_1} \frac{dV}{V}$$

Thus

$$\text{Work done} = C \int_{V_2}^{V_1} \frac{dV}{V}$$

$$= C \log_e \left(\frac{V_1}{V_2} \right)$$

But $V_1/V_2 = P_2/P_1$, therefore,

$$\text{Work done} = P_1 V_1 \log_e \left(\frac{P_2}{P_1} \right)$$

Example 1.6

Calculate the work done if the air in Example 1.5 is compressed isothermally.

Solution

$$P_1 = 1 \text{ bar abs}$$
$$V_1 = 3 \text{ m}^3/\text{min}$$
$$P_2 = 8 \text{ bar abs}$$

$$\text{Work done} = 1 \times 3 \times 10^5 \log_e \left(\frac{8 \text{ N/m}^2}{1 \text{ m}^3/\text{min}} \right)$$

$$= \frac{1 \times 3 \times 10^5 \times 2.08}{60} \text{ N m/s}$$

$$= 10.4 \text{ kW}$$

A much lower value than the polytropic compression calculated in Example 1.5.

Thus, for maximum compressor efficiency the air is cooled during compression. This is accomplished in multi-stage compressors by cooling the air between stages using intercoolers.

1.4.2 Multi-stage compression

Considerable savings in power consumption can be achieved using multi-stage compressors with cooling of the air between stages. In two-stage compressors initial compression takes place in the first or 'low pressure' (LP) stage. Air from this stage is then passed through an intercooler to reduce its temperature back to its initial inlet value, providing cooled air for final compression in the 'high pressure' (HP) cylinder.

The greater the number of stages used for the compression process, the nearer it approaches an isothermal process and the more efficient it becomes. Economic costs limit the number of stages according to the delivery pressure required. At an output pressure of 7 bar, two stages are generally used; at 17 bar it is more usual to have three stages.

Figure 1.11 is a *P–V* diagram for a two-stage compressor with intercooling. The actual compression at each stage is polytropic. If the intercooler pressure is variable then the saving in compressor input power (hatched in the figure) will vary. It can be shown that for minimum input power to a two-stage compressor where the intercooler pressure is P_i, P_1 is the compressor inlet pressure and P_2 is the compressor outlet pressure.

$$P_i = \sqrt{(P_1 \times P_2)}$$

Example 1.7

For a two-stage compressor delivering air at a pressure of 7 bar gauge, determine the intercooler pressure for minimum input power.

Solution

$$P_i = \sqrt{[(7 + 1) \times 1]} = 2.82 \text{ bar abs}$$

$$= 1.82 \text{ bar gauge}$$

1.4.3 Compressor volumetric efficiency

In a piston-type compressor given no mechanical and design limitations there has to be clearance at the end of the compression stroke between the piston head and the cylinder. It is impossible to fully charge the cylinder with air at the end of the suction stroke, and

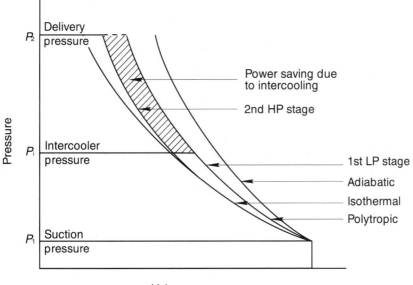

FIGURE 1.11 *P–V* diagram for a two-stage compressor with intercooling.

the pressure in the cylinder will be slightly below atmospheric due to the resistance to flow in the inlet. There may also be some leakage of air through the inlet valve at the changeover from suction to compression, together with leakage across the piston from the high-pressure to the low-pressure side.

Similar conditions exist in all other types of compressors to a greater or lesser extent. Thus the quantity of air delivered by the compressor is less than the theoretical delivery.

$$\text{Volumetric efficiency} = \frac{\text{Volume of f.a.d per minute}}{\text{Swept volume of LP cylinder per minute}}$$

As the efficiency can be affected by the entry conditions of the air, the volumetric efficiency is sometimes quoted at normal 16 °C and 1 atmosphere. (*Note:* Standard temperature and pressure (s.t.p.) is 20 °C and 1 atmosphere).

$$\text{Absolute volumetric efficiency} = \frac{\text{Volume of air delivered per minute at n.t.p}}{\text{Swept volume of the LP cylinder per minute}}$$

When comparing compressors it is essential to work in terms of air delivered, not the displacement or swept volume, as this does not take the volumetric efficiency into account.

Compressor overall efficiency

The theoretical minimum amount of power required to compress the air occurs under isothermal conditions. Consequently, the overall efficiency should be related to isothermal compression.

$$\text{Overall isothermal efficiency} = \frac{\text{Isothermal work done per minute}}{\text{Actual work done per minute}}$$

In practice, depending upon the type of compressor, the value of this efficiency will vary from about 60% to 80%.

1.5 Flow through pipes and pressure drops

Air will be at a temperature well above ambient when it leaves the final stages of a compressor. Cooling takes place in the air receiver and possibly in a heat exchanger or aftercooler. As the air flows through the pipework the system pressure varies, consequently affecting the air temperature according to the type of expansion taking place. Throughout the system the air temperature is constantly changing, making mathematical analysis of the flow and pressure changes extremely complex. By neglecting temperature changes in the system, pressure losses can be approximated using a semi-empirical formula based on steady (non-pulsating flow) through smooth-walled circular pipes:

$$P = \frac{f \times L \times Q^2}{d^5 \times P_m}$$

where P = pressure drop (bar)
f = friction factor (which, for steel pipes used for compressed air at normal temperatures and pressures, may be taken as 500 with the units stated)
L = length of pipe (m)
Q = volume of free air flowing (L/s, i.e. L/s f.a.d.)
d = internal diameter of the pipe (mm)
and P_m = mean or average absolute pressure over the pipe length (bar).

Example 1.8

Estimate the pressure drop over 100 m of pipework of 50 mm bore with a flow rate of 100 l/s. The mean pressure in the pipe may be taken as 5 bar gauge. Take f as 500, then

$$\Delta P = \frac{f.L.Q^2}{d^5.P_m}$$

where f = 500
L = 100 m
Q = 100 l/s
d = 50 mm
P_m = (5 + 1) bar abs

Therefore,

$$\Delta P = \frac{500 \times 100 \times 100^2}{50^5 \times 6}$$

$$= 0.27 \text{ bar}$$

Any pressure drop in the pipework is a loss of energy and, as a consequence, an increase in operating costs. Increasing the bore of the pipe will reduce the pressure drop but increase the cost of the pipe. These must be balanced against each other for optimum conditions. If the pipe bore is increased to 60 mm the pressure drop is given by

$$\Delta P = \frac{500 \times 100 \times 100^2}{60^5 \times 6}$$

$$= 0.107 \text{ bar}$$

Charts and nomographs are available for estimating pressure drops in pipework. When sizing a pipe the possibility of increased flow being needed through that pipe at some time in the future must be considered. The cost of installing pipes and fittings of larger bore sizes than initially required is relatively low when compared with the subsequent cost of altering the installation. Air mains that are too small in diameter will result in high air velocities which prevent water separation. Any water condensate which would normally run as a stream at the bottom of the pipe will be agitated by too high a flow velocity and taken back into the air stream.

In general, the maximum flow velocity in the supply main should be limited to 9 m/s (30 ft/s), but preferably less. When sizing a pipe by this method the quantity flowing must be in terms of the compressed volume.

Example 1.9

A compressor delivers 200 l/s f.a.d. at a pressure of 7 bar. Assuming the maximum flow velocity of 6 m/s, estimate the pipe diameter needed.

Solution

The compressed volume of air flowing will be the free air delivery divided by the compression ratio, where

$$\text{Compression ratio} = \frac{\text{Absolute output pressure}}{\text{Absolute inlet pressure}}$$

Thus, in this case, taking atmospheric pressure as 1 bar,

$$\text{Compression ratio} = \frac{7 + 1}{1}$$

$$= 8$$

$$\text{Flow rate of compressed air} = \frac{200}{8} \text{ l/s}$$

$$= 25 \text{ l/s}$$

Quantity flowing = Area of flow × Average velocity

$$25 \times 10^{-3} \text{ m}^3/\text{s} = \frac{\pi \, d^2}{4} \times 6$$

where d is the bore of the pipe in metres. Thus,

$$d = \sqrt{\left(\frac{25 \times 4 \times 10^{-3}}{6 \times \pi}\right)}$$

$$= 0.073 \text{ m}$$

$$= 73 \text{ mm}$$

In this case an 80 mm bore pipe would be used.

The pressure drop in the pipe can then be calculated or estimated using charts or nomographs, an example of which is shown in Fig. 1.12.

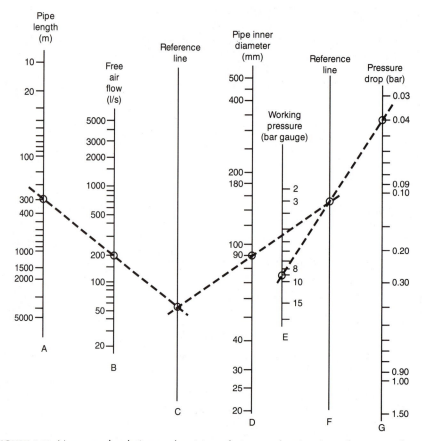

FIGURE 1.12 Nomograph relating to the sizing of pipes and estimation of pressure drops.

Using the nomograph

Determine the approximate pressure drop in a 90 mm diameter, 300 m long pipe if 190 l/s of free air is flowing at a pressure of 9 bar gauge.

Step 1. Locate the pipe length on scale A and the free air flow on scale B. Connect the two points with a straight line and project it through until it intersects the reference line C.

Step 2. Project a line from the intersept point on reference line C through the pipe diameter on scale D until it intersects reference line F.

Step 3. From the working pressure on scale E, project a line through the point on reference line F until it intersects the pressure drop scale G. This point represents an approximate pressure drop of 0.04 bar for the stated conditions.

1.5.1 Pressure drops in pipe fittings

The actual pressure drop occurring in a pipe fitting can only be determined by carrying out practical tests. The results obtained will only apply to that fitting as the internal geometry and surface finish will vary according to the manufacturing process.

Usually the pressure drops in fittings, elbows, trees, etc., in pneumatic circuitry can be neglected. They must, however, be considered when calculating pressure drops in distribution networks. The pressure drop in a fitting is normally expressed in terms of an equivalent pipe length of the nominal bore of the fitting.

Table 1.3 shows typical values.

TABLE 1.3 Pressure drop in pipe fittings expressed as equivalent length in metres of pipe.

Fitting	Nominal pipe size (mm)										
	15	20	25	32	40	50	65	80	100	125	150
90° bend	0.15	0.2	0.25	0.35	0.5	0.6	0.8	1.0	1.2	1.5	1.8
90° elbow	0.25	0.4	0.5	0.65	0.8	1.0	1.4	1.8	2.4	3.2	3.6
Run of tee	0.2	0.3	0.4	0.5	0.7	0.85	1.1	1.3	1.6	2.0	2.5
Side outlet tee	0.5	0.7	1.4	1.8	2.4	2.7	3.6	4.6	5.7	7.0	8.5

CHAPTER 2

Production and distribution of compressed air

2.1 Types of compressor

The majority of industrial compressors are of the positive displacement type, either rotary or reciprocating. Blowers and dynamic type compressors which impart kinetic energy to the air are beyond the scope of this book.

2.1.1 Reciprocating compressors

These consist of one or more pistons reciprocating within a cylinder. The piston may be arranged as a single- or double-action unit. A diagrammatic layout of a two-stage reciprocating compressor with intercooler is shown in Fig. 2.1.

Reciprocating compressors date from the blowing cylinders developed in the eighteenth century for ventilating mine workings. This type of unit has been subjected

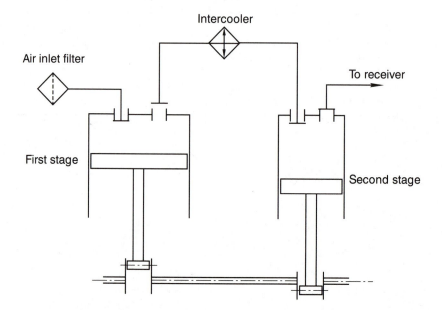

FIGURE 2.1 Diagrammatic representation of a two-stage reciprocating compressor.

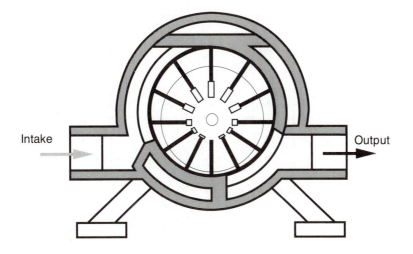

FIGURE 2.2 Diagrammatic representation of a vane compressor.

to continuous development and considerable improvements in design have occurred. Some reciprocating compressors are mechanically unbalanced and require substantial foundations. The 'Vee' type layout can be almost dynamically in balance and thus requires no foundations whatsoever.

2.1.2 Rotary compressors

Vane type compressors

These consist of a cylindrical rotor with radial slots, each carrying a blade or vane. As the rotor spins, centrifugal force keeps the tip of the vane in contact with the compressor body. The inlet and discharge ports are positioned so that the volume between adjacent vanes increases over the suction part of the cycle and decreases over the discharge portion. A diagrammatic layout of a vane-type compressor is shown in Fig. 2.2.

The vanes are lubricated and sealed with oil which is injected during the compression cycle. The injected oil cools the partially compressed air and reduces the volume available for the air, effectively making the unit into a two-stage compressor. The discharge from the compressor contains large amounts of oil which is separated by baffles and filters from the compressed air.

The output air flow is almost pulseless, no air receiver being needed to smooth the delivery. The compressor is nearly perfectly balanced and consequently no foundations are needed.

Screw type

These are based on a Lysholm screw which is in effect a male and a female intermeshing lobe screw. Air is trapped in a cavity between adjacent threads and the casing, the ends of the cavity are sealed by the screws intermeshing. As the screws rotate the air is transferred along the length of the screws from the suction port at one end to the delivery port at the other. A typical screw compressor is shown in Fig. 2.3.

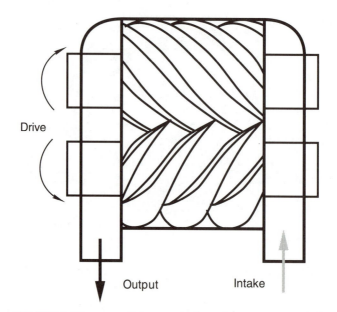

FIGURE 2.3 Diagrammatic representation of a screw compressor.

The continuous delivery action gives a pulse-free supply. Oil is injected into the compressor to seal and lubricate the screws and to cool the air.

Oil-free screw compressors are also generally available, in these types of compressor the two screws have intermeshing gears to drive them. For this reason the screws themselves do not require lubricating, thereby producing an oil-free air supply suitable for many industries.

Screw-type compressors are vibration free and require no foundations.

2.2 **Compressor control**

The output from the compressor has to be matched to the circuit demand which will usually be variable. The compressor output will be in excess of the average demand, and often considerably in excess to allow for future expansion of the system. An air receiver is used as a reservoir of compressed air to average out the system demand. The compressor delivery is fed to the receiver which supplies the system. Any excess air supplied to the receiver could be 'blown off' through the safety or maximum pressure valve to atmosphere, but this would be a very expensive and impractical method.

The air delivered by the compressor can be matched to the system demand by one of the following control methods.

  1. Continuously varying the amount of air entering the compressor by restricting or throttling the air inlet to the compressor.
  2. Continuously varying the drive speed of the compressor.

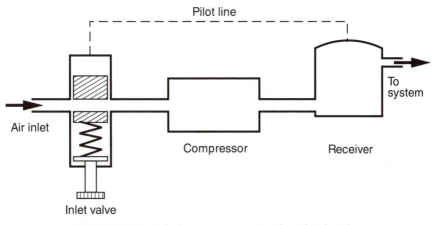

FIGURE 2.4 Control of compressor output by inlet throttling.

3. Lifting the air inlet valves to unload the compressor.
4. Stopping and starting the drive to the compressor.

Each of these methods has applications and relative merits.

2.2.1 Throttling

A pilot-operated valve controls the quantity of air admitted to the inlet side of the compressor. This is shown in Fig. 2.4.

The inlet valve is controlled by a pilot line connected to the receiver. When the pressure exceeds the pre-tension in the valve control spring the valve starts to close; the higher the pilot pressure, the higher the restriction in the suction line. Although the throttle valve limits the maximum receiver pressure it is still essential to have a safety valve fitted to any receiver.

It should be noted that when a large compressor is running off load or idling it may require 20 per cent of full load power. A small compressor may require an even higher percentage. This energy will mainly appear as heat within the unit and has to be dissipated by the cooling system.

2.2.2 Variable speed drive

This method can be very efficient on certain types of compressor, such as those without a minimum speed limitation, but it can be costly to effect. A variable speed electric motor drive requires an expensive control system whether it is an a.c. or d.c. motor. An internal combustion engine has a limited speed range, and although suitable for mobile applications it is not ideal for use within a factory. There are several types of variable speed drives from hydrostatic transmission to mechanical drives, all of which are relatively expensive. Consequently, this method of compressor control is not very widely used.

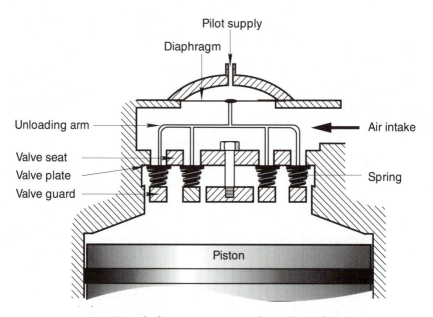

FIGURE 2.5 Control of compressor output by suction valve opening.

2.2.3 Unloading

The compressor is unloaded by keeping the suction valve open (see Fig. 2.5) so that air passes through the suction pipework in one direction during the suction stroke and in the reverse direction during the compression stroke.

This type of control can only be applied to reciprocating compressors and has the disadvantage that the unit is noisier off-load than on-load. The suction pipe design and the use of silencers will reduce the noise level.

2.2.4 Stopping and starting

This involves discontinuing the drive to the compressor and can be achieved by:

(a) the use of an automatic Star-Delta starter for an electric motor drive, but it should be noted that the maximum number of starts per hour should not exceed 20 or the electrical control gear is subject to excessive wear;

(b) an automatic clutch unit which disengages the compressor drive, but again the number of starts per hour should be limited to reduce the clutch wear.

The size of the air receiver and compressor determines the number of starts per hour. When the capacity of the compressor is twice the circuit demand, the highest number of starts per hour will occur and the compressor will spend half its time 'on load' and half 'off load'. The larger the effective capacity of the receiver – i.e. capacity of receiver plus associated pipework – the lower the number of starts per hour.

Example 2.1

An air compressor delivers 10 m³/min f.a.d. at a pressure of 7 bar. The average circuit demand being 7 m³/min f.a.d. with an allowable fluctuation of delivery pressure of from 7 to 6 bar. Assuming that a receiver having a capacity of 4 m³ is used, determine the number of times the compressor comes on load per hour. Assume the air is fully cooled and its temperature is constant. All pressures are gauge, and atmospheric pressure is 1 bar absolute.

Solution

To find the quantity of free air stored in the air receiver apply the characteristic gas equation:

(a) at 7 bar gauge

$$\frac{P_1 V_1}{T_1} = \frac{P_2 V_2}{T_2}$$

$$P_1 = 1 \text{ bar abs} \qquad V_1 = ?$$

$$P_2 = (7 + 1) \text{ bar abs} \qquad V_2 = 4 \text{ m}^3$$

$$T_1 = T_2 \text{ (constant temperature)}$$

$$V_1 = \frac{P_2 V_2}{P_1} = \frac{8 \times 4}{1}$$

$$= 32 \text{ m}^3$$

(b) similarly, at 6 bar gauge

$$\frac{P_3 V_3}{T_3} = \frac{P_4 V_4}{T_4}$$

$$P_3 = 1 \text{ bar abs} \qquad V_3 = 4 \text{ m}^3$$

$$P_4 = (6 + 1) \text{ bar abs} \qquad V_4 = ?$$

$$V_4 = \frac{P_3 V_3}{P_4} = \frac{7 \times 4}{1}$$

$$= 28 \text{ m}^3$$

The difference in the volume of free air stored in the receiver between 7 bar and 6 bar pressure is 4 m³. The compressor will come on load when the receiver pressure falls to 6 bar and go to off load when it reaches 7 bar. The compressor delivers 10 m³/min f.a.d. and the circuit requires 7 m³/min f.a.d. This leaves 3 m³/min to charge the receiver. Thus,

$$\text{Charge time} = \frac{4 \text{ m}^3}{3 \text{ m}^3/\text{min}}$$

$$= 1.33 \text{ min}$$

The receiver discharges from 7 bar to 6 bar gauge supplying the system. There is a 4 m³ f.a.d. during this time. Thus,

$$\text{Discharge time} = \frac{\text{Air available from receiver}}{\text{Air required/min by system}}$$

$$= \frac{4 \text{ m}^3}{7 \text{ m}^3/\text{min}}$$

$$= 0.57 \text{ min}$$

Thus the total time between the compressor going on load is:

$$1.33 + 0.57 = 1.9 \text{ min}$$

This represents 31.6 starts per hour, which is too many, so the receiver size must be increased to, say, 6 m³. This will increase the quantity of air stored between the pressures of 6 and 7 bar gauge to 6 m³.

So

$$\text{Charge time} = \frac{6 \text{ m}^3}{3 \text{ m}^3/\text{min}}$$

$$= 2 \text{ min}$$

$$\text{Discharge time} = \frac{6 \text{ m}^3}{7 \text{ m}^3/\text{min}}$$

$$= 0.86 \text{ min}$$

This gives a cycle time of 2.86 min, which is acceptable. The cycle time can be increased by either increasing the capacity of the receiver or increasing the range of pressures over which the system can function.

2.3 Air conditioning

As already mentioned, atmospheric air contains moisture and particles of dirt. When the air leaves the compressor it will contain contaminates generated by the compressor, including burnt compressor oil which will combine with the moisture and dirt present to form a thick gummy substance which can cause components to malfunction.

When steel pipework is used, rust forms due to moisture in the air; even galvanised pipe will rust where there are pipe joints. Scale will eventually break off the pipework and enter the air flow.

The air receiver, when one is fitted to the system, will act as a settling volume, where droplets of oil and water together with the largest dirt particles can separate out and fall to the bottom of the receiver. A manual or automatic drain must be fitted into each receiver and regularly checked to ensure that any liquid collected is drained off. A gate or ball valve can be used as a manual drain but it must always be carefully closed after it has operated to prevent any air leaks. Automatic drain valves will have some type of float shutting off the drain port; when the liquid level rises in the valve body the float

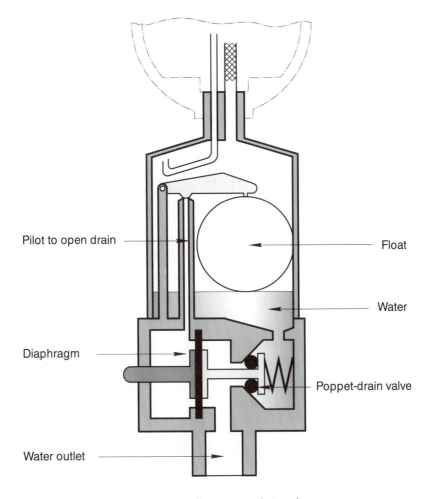

Pilot to open drain

Float

Water

Diaphragm

Poppet-drain valve

Water outlet

FIGURE 2.6 (a) Ball type auto-drain valves.

rises and opens the drain port. A diagrammatic sketch of a ball-type drain valve is shown in Fig. 2.6(a).

The auto-drain valve shown in Fig. 2.6(b) has a volume of liquid trapped below the level of the discharge port. This allows any large particles of scale to settle into this 'dead' volume and prevents the discharge valve from being jammed and not being able to close fully.

A separate manual valve is fitted to enable this 'dead' volume to be emptied periodically, the system air pressure blowing the fluid and dirt through the valve. On some types of auto-drain, a secondary poppet valve is fitted to block the inlet to the valve when the float lifts. This shuts off the main air supply and prevents the drained fluid being blasted out when the valve operates.

Auto-drains are available in different sizes for different applications. They can be fitted onto the main air receiver, the drain legs in the supply pipework or ring main and onto water traps and filters.

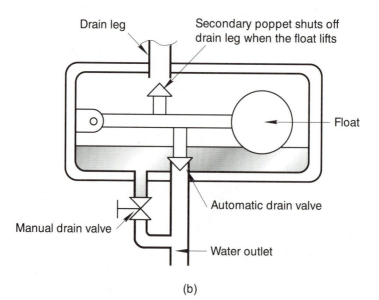

(b)

FIGURE 2.6 (b) Ball type auto-drain valves.

Where quantities of water are removed from pneumatic systems care must be taken to dispose of the condensate in a manner acceptable to the local bye-laws. Since the condesate is likely to contain oil, it is not generally acceptable to allow this to enter the ground-water system.

Water removal

The air leaving the compressor will contain droplets of water and be at a temperature considerably above the ambient. The air under these conditions is fully saturated and a reduction in its temperature will cause moisture to condense out. Some of this free moisture will separate out in the air receiver and pipework and can be removed by drain valves. However, some of the smaller droplets of water will be carried by the stream of air to the components, with further condensation occurring as the compressed air cools.

In applications such as instrumentation or precision control systems it is imperative that the air used be clean and dry with a reduced dew point. This is achieved by drying the air at a high pressure and then reducing the pressure by expanding the air, so lowering the dew point.

2.3.1 Air driers

There are three common methods used to dry air: adsorption, absorption and refrigeration.

Adsorption

Adsorption-type driers consist of two pressure chambers filled with a water-adsorbing chemical. The air to be dried is passed through one chamber and the moisture removed, while the other chamber containing spent chemical is being regenerated by either heating the chemical or passing hot dry air over it. So at any time one chamber is drying the air and the other is being dried.

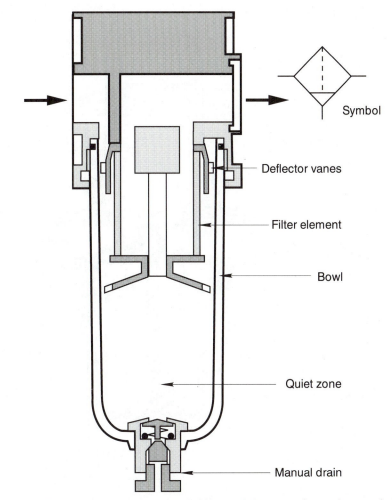

Symbol

Deflector vanes

Filter element

Bowl

Quiet zone

Manual drain

FIGURE 2.7 General arrangement of a filter and drain together with its symbol.

Absorption
Absorption driers consist of a chamber containing a chemical which absorbs the moisture, forming a solution which drains to the base of the chamber, where it is drained and disposed of. The absorption-type drier has to be periodically drained and recharged with chemicals.

Refrigeration
Refrigerant driers cool the air down to 1 °C or below, thereby causing precipitation of moisture which is separated out using auto-drain valves. The refrigerant driers are heat exchangers which cool the air, reducing its ability to retain moisture and, therefore, causing precipitation which is drained from the drier. The air is then reheated to the original temperature thereby producing relatively dry air. All types of driers incur running costs and so the requirement for this type of conditioning of air for general use needs careful consideration.

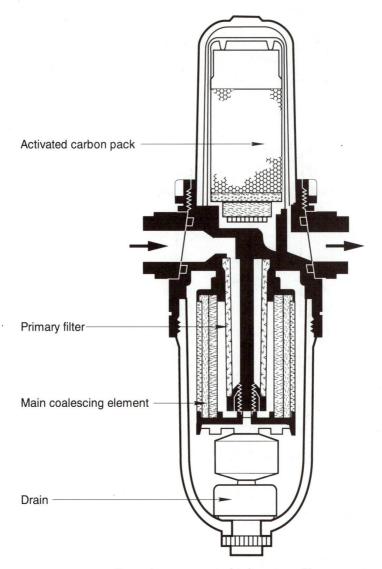

Activated carbon pack

Primary filter

Main coalescing element

Drain

FIGURE 2.8 General arrangement of a three-stage filter.

2.3.2 Air filters

Air line filters are used to remove free droplets of liquid, water and oil together with particles of dirt. A typical filter is shown in Fig. 2.7.

As the air enters the filter it flows over blades which cause the air to spin. This spinning motion throws the droplets of liquid and larger dirt particles outwards by centrifugal force onto the filter bowl, from which they run down to the bottom of the bowl below the baffle into the quiet zone. A manual or automatic drain removes the collected liquid. The air now flows through a filter element consisting of a block of porous material which may be sintered brass, stainless steel or cellulose.

An air filter is rated in terms of the largest spherical particle – measured in micrometres (10^{-6} m), generally referred to as microns – that will pass through the filter and the volume of air per minute that it can filter efficiently. When the air is to be used for instrumentation, air gauging, paint spraying, air bearings, breathing air and similar applications, it must be oil free. This can be achieved by either using an oil-free compressor (a carbon ring, diaphragm-type or screw-type compressor) or by removing all traces of oil from the air using a two- or three-stage filter as shown in Fig. 2.8.

This filter comprises three sections: in the first section the large contaminants are removed by the sintered pre-filter medium; the second stage is coalescing filter element covered by a porous plastic sock to prevent re-entrainment of separated liquids; and the third stage consists of an activated carbon element which adsorbs some hydrocarbon gases which are caused by decomposition of the compressor oil. Should there be a malfunction in the coalescing filter the third stage will also absorb the oil carry-over. It should be noted, however, that the oil contamination of this section will result in the inability of this part of the filter to adsorb hydrocarbons effectively. Should there be a change of colour of this third section, then the filter requires servicing.

Filter selection and maintenance

A general-purpose 5-micron oil-removing single-stage filter is adequate for the vast majority of pneumatic applications, with activated carbon filters being used where breathing air is required.

The filter capacity is usually given in terms of free air flow and must be carefully matched to the system requirements. An oversized filter must **not** be used, as it would reduce the air velocity through the centrifugal section and, consequently, lower the efficiency of moisture and dirt separation. Too small a filter results in high pressure drops, high air flow velocities and re-entrainment of water in the air stream.

Filters and moisture separators must receive regular attention; the filter bowl on manual drain filters must be emptied and cleaned on a regular basis. Automatic drain filters require regular cleaning of valves and valve seats. The filter element must be changed or cleaned in accordance with manufacturers' recommendations. The filter unit should be fitted immediately upstream of the point at which the air is to be used.

2.3.3 Air regulators

A regulator should be used before any pneumatic system to set the air pressure to the minimum value at which the pneumatic system operates satisfactorily. The regulator reduces the air pressure to the required value, and the air expands as the pressure is reduced. This results in a saving in air consumption, giving a more efficient system. A general-purpose regulator, together with its symbol, is shown in Fig. 2.9.

The valve spool position is adjusted by a control spring acting through a diaphragm to open the valve. The downstream pressure is fed to the bottom of the diaphragm and acts against the control spring. When the desired pressure setting is reached, the valve spools lifts under the action of the return spring; this throttles the air and reduces the pressure. If there is no demand from the system the regulator spool closes; but if demand

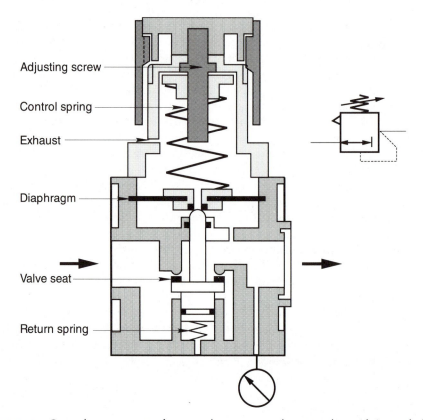

FIGURE 2.9 General arrangement of a general-purpose regulator together with its symbol.

increases, the spool opens just sufficiently for the regulated pressure to balance the load set by the adjustable spring. If there is any tendency for the system to build up a back pressure, which could be undesirable, a relieving-type regulator should be used. This is similar to the standard regulator but has an air bleed hole in the valve seat on the diaphragm. Any excessive pressure buildup is vented to the atmosphere through this vent hole.

This type of regulator should be used only when non-precision regulation is required as the regulated pressure depends to a certain extent on the volume of the air flowing through the regulator. A typical curve is shown in Fig. 2.10.

The exact characteristics should be checked with the manufacturer if there is any doubt as to the suitability of any particular regulator. Where precision control of air pressure is needed for instrumentation or control applications, a multi-stage regulator must be used. A section through a multi-stage precision regulator is shown in Fig. 2.11.

This type of unit is only suitable for clean, oil-free air. The pilot section sets the downstream pressure to the required value; the pilot diaphragm is in a force balance condition due to the action of the adjustable spring force and the regulated pressure. Any variation in downstream pressure causes a movement of the pilot section opening or

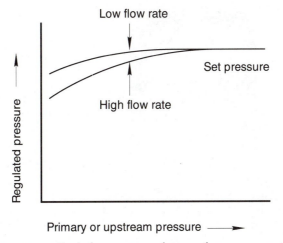

FIGURE 2.10 Typical pressure regulator performance curves.

closing the bleed orifice from the main section. This gives a very rapid response of the main or power section opening or closing the supply valve. The valve shown has a relieving feature to prevent any buildup of downstream pressure. There is always a slight air bleed through the exhaust port. This air bleed is necessary to give the rapid response and high accuracy of this type of valve.

Whatever type of regulator is used it must be positioned as near as possible to the point of application of the regulated air.

2.3.4 Air lubricators

The air from the compressor has been first passed through a filter unit to remove dirt, water and oil droplets, and is then passed through a regulator to set the correct pressure. In a large number of applications oil has to be injected into the supply to give adequate lubrication to the valves, cylinders and motors in the system. An air line lubricator injects very fine particles or droplets of clean oil into the air stream.

Two types of lubricator are in general use: the 'fog' type and the 'mist' type, which is sometimes referred to as the 'micro-fog' type. In the oil fog lubricator shown in Fig. 2.12, the air passing through the throat creates a venturi effect, drawing oil from the reservoir up the siphon tube. The oil then enters the sight-feed dome where it is controlled by a metering screw, dripping at the prescribed amount into the air flow path where it is atomised.

The atomised oil forms an airborne fog of varied particle size, generally in excess of 2 microns. This fog is carried in the air stream for a distance up to a maximum of about 10 m (33 ft). Oil fog lubricators should not be expected to provide adequate lubrication over greater lengths of pipe run than 10 m. They should not, in general, be used to lubricate more than one appliance per system.

In the micro-fog or oil mist lubricator shown in Fig. 2.13, air is directed into the oil reservoir. This creates a flow of oil into the sight dome via a metering needle valve

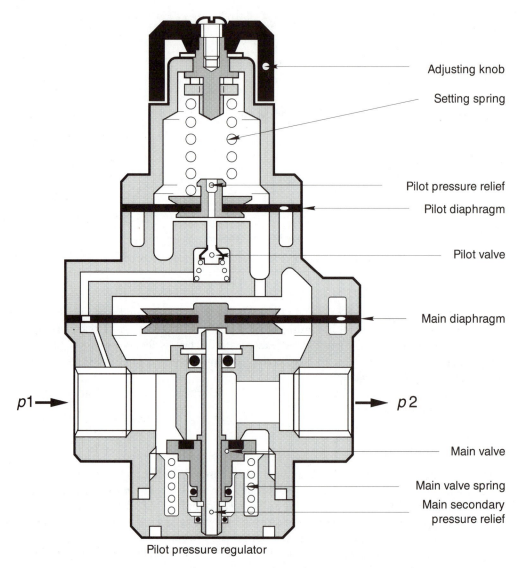

Adjusting knob

Setting spring

Pilot pressure relief

Pilot diaphragm

Pilot valve

Main diaphragm

*p*1

*p*2

Main valve

Main valve spring

Main secondary
pressure relief

Pilot pressure regulator

FIGURE 2.11 General arrangement of a multi-stage pressure regulator.

and then through the atomising jet and back into the upper part of the bowl. Here oil particles greater than about 2 microns fall out, returning to the oil reservoir. The small particles of oil remaining airborne can be carried out of the lubricator and through long distances without separating out.

Siting and sizing lubricators

Any lubricator requires a minimum air flow rate at any given pressure for it to work correctly, so it is essential that the manufacturer's data be consulted. The type of oil used

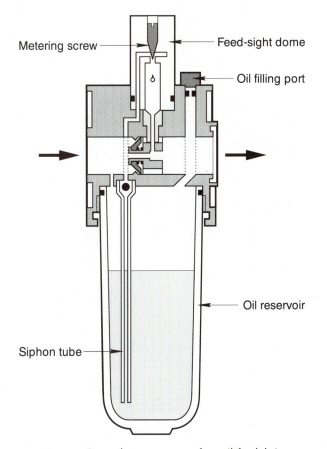

Metering screw

Feed-sight dome

Oil filling port

Oil reservoir

Siphon tube

FIGURE 2.12 General arrangement of an oil fog lubricator.

in the lubricator must be compatible with the system components, and it is usual to use a good quality hydraulic oil. The lubricator must also be positioned as near as possible to the unit it is serving. The lubricator should be adjusted to give the correct air/oil ratio, as excessive oil can cause malfunction of components and be dripped from exhaust ports of valves, contaminating the surrounding environment.

2.4 Compressor plant layout

In order to reduce pipe runs, the compressor installation should be located as closely as possible to the point at which the air is to be used. The air intake to the compressors must be as cool, clean and dry as possible to increase the efficiency and reduce the conditioning costs. The running noise of the compressor must also be considered; it should be located to cause minimum nuisance, as sound-proofing can be very costly and sometimes not very successful. A schematic layout of a typical compressor installation is shown in Fig. 2.14.

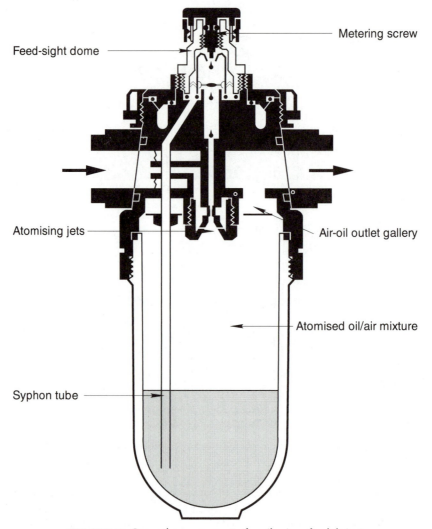

Metering screw

Feed-sight dome

Atomising jets

Air-oil outlet gallery

Atomised oil/air mixture

Syphon tube

FIGURE 2.13 General arrangement of a oil micro-fog lubricator.

The compressor intake from outside the plant should be situated in a cool area with a canopy or hood over the intake to protect it from rain. This intake should be sited as high up as possible so as to avoid the majority of airborne contaminants. Care should be taken to avoid steam vents, air-conditioning vents, internal combustion exhausts and chemical fume outlets.

An inlet filter, which can be of wire wool dampened with mineral oil, is used to remove as much of the airborne dirt as possible. A baffle-type silencer can be installed just prior to the compressor, which is particularly important if compressor control is to be achieved by lifting the suction valve.

In multi-stage compressors the intercooler between stages will be either air blast or water type, as will the aftercooler. The effect of cooling the air causes water vapour

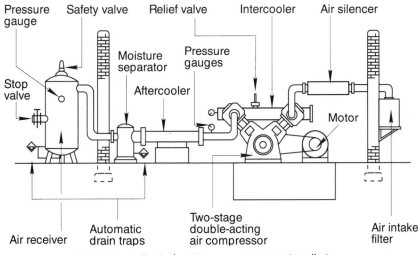

Pressure gauge Safety valve Relief valve Intercooler Air silencer

Moisture separator Pressure gauges

Stop valve Aftercooler Motor

Air receiver Automatic drain traps Two-stage double-acting air compressor Air intake filter

FIGURE 2.14 Typical stationary compressor installation.

to condense out; this is removed at the drains, which may be automatic or manual. The cooled air is fed into the air receiver which is best located outside the compressor room in a cool area. As the air in the receiver will cool further, a drain has to be fitted to the receiver to remove the condensate. In lubricated compressors there will be some oil carry-over in the air; however, this oil will be contaminated or oxidised and be unsuitable for the lubrication of pneumatic components. The compressor oil, therefore, has to be removed from the air, and this will happen to some degree in the receiver. The longer the air stays in the receiver (up to a limit), the more oil and water will separate out.

2.5 Air line installation

Sizing the air line
As the air flows through the pipework there is a pressure drop, the value of which depends on the diameter, length of the pipe, and the quantity of air flowing. This pressure drop represents wasted energy and, as such, should be kept to a minimum. A guide to the relationship between pipe bore and flow rate for a pressure drop of 1 bar per 100 m of pipe is given in Table 2.1.

Air line layout
The air line may be constructed of steel pipe, preferably galvanised, or of rigid plastic pipe. In the case of plastic pipes high-temperature areas must be avoided. The pipework is best laid out as a ring main (Fig. 2.15) to give two flow routes to every take-off point.

The pipework should slope at about 1 in 100 to allow any condensate to run to a drain point. Branch lines and machine supply lines must come from the top of the ring main pipe to prevent the ingress of condensate. The compressors should be connected to the receiver or ring main by shut-off valves to enable a compressor to be taken out of service for maintenance.

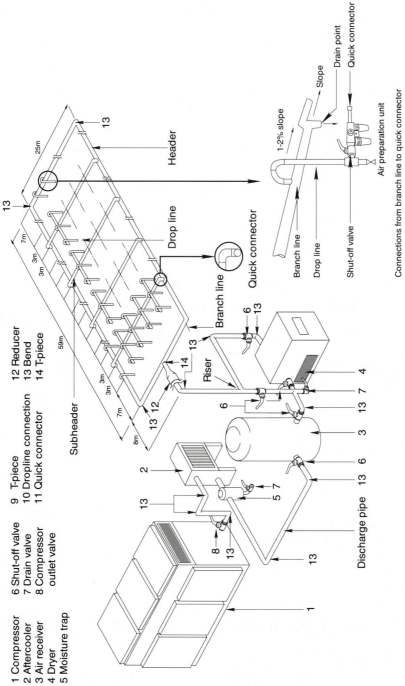

1 Compressor   6 Shut-off valve   9 T-piece
2 Aftercooler   7 Drain valve   10 Dropline connection
3 Air receiver   8 Compressor   11 Quick connector
4 Dryer   outlet valve   12 Reducer
5 Moisture trap   13 Bend
                                        14 T-piece

FIGURE 2.15 Typical ring main system.

TABLE 2.1 Typical airflow rates l/s through metric pipe at various pressures.

Nominal bore (mm)	Gauge pressure (bar)			
	7	8	9	10
20	5.4	5.8	6.2	6.5
25	9.4	10.1	10.7	11.4
32	20.4	21.9	23.4	24.7
40	30.9	33.3	35.4	37.5
50	49.6	53.3	56.8	60.1
65	100.7	108.2	115.3	122.1
80	182.5	196.2	209.1	221.3
100	368.0	395.6	421.6	446.3
120	655.5	704.6	750.9	794.9
150	1054.5	1133.4	1207.9	1278.7
200	2245.0	2413.0	2572.0	2722.0

2.6 Air consumption

The quantity of air consumed by an installation can be one of the major running costs of the plant. A popular misconception is that compressed air is free. On the contrary, a compressor producing 75 l/min of free air at a pressure of 7 bar gauge will require an input power of about 1 kW.

Alternatively, to produce a flow of 6 ft³/min of free air at 100 lb/in² will require an input power to the compressor of about 1 kW. These values apply to smaller compressors of up to about 50 kW input; the efficiency of larger units will increase slightly but the actual cost of the compressed air is still considerable.

It is usual to express air consumption in terms of free air delivered (f.a.d.). This is the volume a given quantity of compressed air would occupy at atmospheric pressure and at the same temperature. Expressing the air demand of systems and compressor deliveries in terms of f.a.d. simplifies calculations and enables comparisons to be made between compressors or systems operating at different pressures.

Example 2.2

A double-acting pneumatic cylinder with a bore of 100 mm, a rod diameter of 32 mm and a stroke of 300 mm operates at a pressure of 6 bar gauge on both extend and retract strokes. If the cylinder makes 25 complete cycles per minute calculate the air consumption.

Solution

Calculate the swept volume extending and retracting.

$$\text{Extend} = \frac{\pi \times 100^2 \times 300}{4} \text{ mm}^3$$

(*Note*: $1 \text{ mm}^3 = 1 \text{ m}^3 \times 10^{-9}$ or $1 \text{ mm}^3 = 1 \text{ litre} \times 10^{-6}$.) Therefore,

$$\text{Extend volume} = \frac{\pi \times 100^2 \times 300}{4} \text{ mm}^3$$

$$= 2.355 \text{ litres}$$

$$\text{Retract volume} = \frac{\pi(100^2 - 32^2)}{4} \times 300 \text{ mm}^3$$

$$= 2.114 \text{ litres}$$

Total volume of compressed air per cycle is

$$2.355 + 2.114 = 4.469 \text{ litres}$$

$$\text{Volume of compressed air per minute} = \text{volume per cycle}$$

$$\times \text{ Cycles/min}$$

$$= 4.469 \times 25$$

$$= 111.7 \text{ l/min}$$

To express this in terms of free air, the volume the air would occupy at atmospheric pressure has to be calculated, taking atmospheric pressure as 1 bar absolute.

Then, by the gas law,

$$\frac{P_1 V_1}{T_1} = \frac{P_2 V_2}{T_2}$$

Assuming isothermal conditions and expressing the pressure as an absolute value, then

$$P_1 = 1 \text{ bar abs}$$

$$V_1 = \text{Air consumption f.a.d.}$$

$$P_2 = 6 \text{ bar gauge}$$

$$= (6 + 1) \text{ bar abs}$$

$$V_2 = 111.7 \text{ l/min}$$

and

$$T_1 = T_2$$

Substituting in $P_1 V_1 = P_2 V_2$

$$V_1 = (6 + 1) \times 111.7$$

$$V_1 = 782 \text{ l/min f.a.d.}$$

If more convenient, this can be converted into cubic feet per minute.

Note: 1 ft³ = 28.6 litres, so

$$V_2 = \frac{782}{28.6} = 27.4 \text{ ft}^3/\text{min f.a.d.}$$

If the extend stroke of the cylinder is the work stroke, the air pressure needed to give sufficient thrust for the extend stroke usually sets the pressure used for retracting the cylinder. This may be very much higher than is actually needed. If the air used to retract the cylinder can be supplied at a lower pressure, this will reduce the air consumption.

A method of supplying the cylinder with different air pressures for the extend and retract stroke is shown in Fig. 2.16.

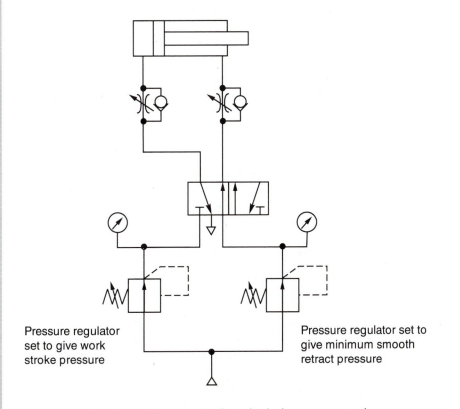

Pressure regulator set to give work stroke pressure

Pressure regulator set to give minimum smooth retract pressure

FIGURE 2.16 Energy saving by individual pressure control.

Example 2.3

Calculate the percentage reduction in air consumption per minute for the cylinder in Example 2.2, if the supply pressure used during the retract stroke is 2 bar gauge.

Solution

From previous calculation:

$$\text{Retract volume per stroke} = 2.114 \text{ litres}$$

The retract stroke is at 2 bar gauge, so volume of free air per retract stroke is given by

$$P_1V_1 = P_2V_2$$

or

$$V_1 = \frac{P_2V_2}{P_1} = \left(\frac{2+1}{1}\right) \times 2.114$$

$$= 6.34 \text{ litres f.a.d.}$$

From previous calculation:

$$\text{Extend volume per stroke} = 2.355$$

The extend stroke is at 6 bar gauge, so volume of free air for the extend stroke is obtained using the formula

$$P_1V_1 = P_2V_2$$

or

$$V_1 = \frac{P_2V_2}{P_1}$$

$$\text{Extend volume} = \frac{(6+1)}{1} \times 2.355$$

$$= 18.48 \text{ litres f.a.d.}$$

Total air consumption per cycle is

$$18.48 + 6.34 = 24.82 \text{ litres f.a.d.}$$

At 25 cycles/min the air consumption is

$$24.8 \times 25 = 620 \text{ l/min f.a.d.}$$

The air consumption calculated in Example 2.2 – when the extending and retracting pressures were 6 bar gauge – was 782 l/min f.a.d. By reducing the retract pressure to 2 bar gauge the air consumption was reduced to 620 l/min f.a.d. The percentage reduction is

$$\frac{782 - 620}{782 \times 100} = 20.7\%$$

So far only the amount of air used in actually operating the cylinder has been considered. An additional amount is used in pressurising the pipework between the valve and the cylinder. Since it is common practice to use pipework having the same or similar bore to the port sizes of the valve, this will give satisfactory operational cylinder speeds if the valve has been correctly sized. There will, however, be a time delay between the valve being operated and the start of movement of the piston, the delay is partly due to the time taken to pressurise the volume of pipework between the valve and cylinder – the less the volume, the less the delay. The volume of the pipe is dependent upon the pipe bore and its length, and to reduce this volume the valve should be located as closely as possible to the cylinder.

Example 2.4

Consider the cylinder details given in Example 2.2. Calculate the increase in air consumption when the pipework between the valve and cylinder is taken into account. The bore of the pipe is 12 mm and the distance between the cylinder and the valve is 1 m.

Solution

$$\text{Volume of pipework} = \pi \times \frac{12^2}{4} \times 1000 \times 2 \text{ mm}^3$$

Note there are two pipes

$$= 226{,}224 \text{ mm}^3$$

$$= 0.226 \text{ litre}$$

Both pipes are pressurised to 6 bar gauge once per cycle or 25 times per minute.

Note: one pipe is pressurised on the extend stroke; the other pipe is pressurised on the retract stroke. Thus the volume of air used in pressurising the pipework is:

$$0.226 \times \left(\frac{6+1}{1}\right) \times 25 = 39.6 \text{ l/min f.a.d.}$$

For Example 2.2 the air consumption for the cylinder alone is 782 l/min f.a.d. The air 'wasted' in the pipework adds another 40 l/min or 5 per cent to this figure.

The air consumption of a pneumatic system may be considerably reduced by using pressure regulators to set the air pressure to the minimum value for correct functioning of the cylinder in both extend and retract mode, and by locating the control valve as near to the actuator as possible.

In large multi-cylinder installations it is often economical to have dual pressures, one to power the work stroke the other to return the actuators. The cylinders being connected are shown in Fig. 2.17.

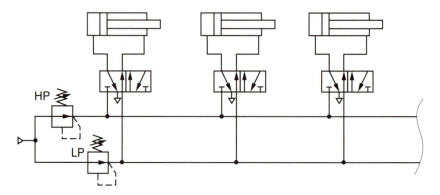

FIGURE 2.17 Energy saving by dual pressure control of multi-cylinder circuits.

2.6.1 Determination of compressor capacity

In order to estimate the size of compressor installation needed, an educated guess has to be made into the future uses of compressed air within the plant. The theoretical consumption of existing plant can be calculated when it involves machines whose duty cycles are fully known, but many pneumatic devices are used intermittently – for example, air guns used to blow off swarf from machined components. As a general rule of thumb, calculate the air consumption of machines in the plant and then double this value; if future expansion is considered, it would be as well to treble the estimated air demand.

When selecting the compressors, the ideal would be to have three compressors of equal capacity, two of which would be running to fully supply the systems demand. The third unit would be on standby, so that one compressor could be taken out of service to allow maintenance or repair work to be carried out without interrupting production.

When several compressors are employed to supply air via a common receiver, the compressor control is important to optimise the system. If a method of unloading and loading is used, then compressors should come on and off load at various pressures. Take, for example, a plant which is supplied by three compressors, a fourth unit being on standby. The first compressor could be arranged to cut off at 7.5 bar and back on at 7 bar, the second cut off at 7 bar back on at 6.5 bar and the third cut off at 6.5 bar and back on at 6 bar. Thus, below 6 bar all three compressors would be working and as the pressure increases the compressors would in turn cut out. The unit with the highest pressure setting would do the most work and, consequently, require most maintenance. In order to even out the wear rates the compressor working at the highest pressure will be taken out of service, the replacement compressor will then work at the lowest pressure setting and the remaining two compressors will each have their working range stepped up. If the pressures at which the compressors come on and off load are too close together, then 'hunting' will occur.

It is always preferable to overestimate the air demand and to install larger compressor units than initially needed as the compressor control unit will switch the compressors on and off load as required.

2.6.2 Air receiver sizing

The functions of the air receiver in a pneumatic system depend upon its siting. The receiver located at the compressor installation allows the air to cool after it has been compressed, thus letting the moisture condense out. It also acts as a storage vessel, reducing pressure fluctuations in the air supply when the compressor cuts on and off load. It may also supply air to the system for a time if there is a breakdown of the compressor.

The size of the air receiver and allowable variations in the supply pressure will determine the frequency with which the compressor switches on and off load. The following example illustrates the method for sizing a receiver, assuming the system demand is constant and that only one compressor is used.

Example 2.5

A pneumatic ring main supplies a plant with an average demand of 20 m^3/min f.a.d. The minimum working pressure at the receiver is 5 bar gauge. The air compressor has a rated delivery of 35 m^3/min f.a.d. at a working pressure of 7 bar gauge. The control system switches the compressor off load when the receiver pressure reaches 7 bar gauge rising, and back on load when the receiver pressure is 5 bar gauge falling. If the maximum allowable number of starts per hour of the compressor is 20, determine a suitable receiver capacity.

Solution

As the compressor is limited to 20 starts per hour, the minimum time between starts is 3 minutes, assuming steady operating conditions.

In 3 minutes the system demand is 3 × 20 m^3 f.a.d. To supply this the air compressor must run for

$$\frac{3 \times 20}{35} = 1.714 \text{ min}$$

Thus the compressor is off load for 1.286 min during which time the receiver has to supply the air required by the system while the pressure in the receiver falls from 7 to 5 bar gauge. Therefore,

$$\text{Volume of air supplied from receiver} = 1.286 \times 20$$

$$= 25.72 \text{ m}^3 \text{ f.a.d.}$$

Let V be the actual volume of the receiver; the volume of free air stored in the receiver at 7 and 5 bar gauge can be calculated and equated to the required volume. Assume that the air in the receiver is at constant temperature or that any change in temperature is too small to significantly affect the calculations.

$$\text{Volume of free air in receiver at 7 bar} = V \times \left(\frac{7 + 1}{1}\right) \text{ m}^3$$

$$= 8V \text{ m}^3$$

$$\text{Volume of free air in receiver at 5 bar} = V \times \left(\frac{5+1}{1}\right) \text{ m}^3$$

$$= 6V \text{ m}^3$$

So, volume of free air available from the receiver as the pressure falls from 7 bar to 5 bar gauge is equal to

$$(8V - 6V) = 2V \text{ m}^3$$

The volume of air to be supplied is, as calculated previously, 25.72 m³. Hence

$$2V = 25.72 \text{ m}^3$$

$$V = 12.86 \text{ m}^3$$

Thus the volume of the receiver required is 12.86 m³.

Select a 13 m³ receiver. Some of the volume of the pipework in the ring main can be considered as part of the receiver, and so increase the effective capacity of the receiver.

A situation which often occurs is when the system demand changes considerably from the value on which the compressor plant and air receiver were designed. A reduction in air demand, or an increased demand, can cause problems in compressor control. These problems can easily be understood by considering a particular case.

Example 2.6

The air demand from the plant in Example 2.5 has reduced to 15 m³/min f.a.d. The same compressor of 35 m³/min f.a.d. at 7 bar gauge together with a 13 m³ receiver are used. Calculate the number of starts per hour of the compressor if the same switching pressures of 7 bar and 5 bar gauge are used.

Solution

Volume of air stored in receiver at 7 bar is

$$13 \times \left(\frac{7+1}{1}\right) = 104 \text{ m}^3 \text{ f.a.d.}$$

Volume of air stored in receiver at 5 bar is

$$13 \times \left(\frac{5+1}{1}\right) = 78 \text{ m}^3 \text{ f.a.d.}$$

Therefore, the volume of air available from receiver between 7 bar and 5 bar pressure is

$$104 - 78 = 26 \text{ m}^3 \text{ f.a.d.}$$

The system demand is 15 m³/min f.a.d., thus the air stored in the receiver will run the system for 26/15 or 1.734 min.

When the receiver is being charged the volume of air entering the receiver is the difference between the compressor output and the system demand. In this case:

$$35 - 15 = 20 \text{ m}^3 \text{ f.a.d. of air flow into the receiver}$$

Thus the time taken to charge the receiver is taken for 20 m³ f.a.d. of air flow into the receiver. Thus,

$$\text{Charge time} = \frac{26}{20}$$

$$= 1.3 \text{ min.}$$

The cycle time for the receiver, i.e.

$$\text{Charge plus discharge time} = 1.73 + 1.3$$

$$= 3.03 \text{ min}$$

This gives the number of starts per hour as

$$\frac{60}{3.03} = 19.9$$

This is less than the suggested 20 starts per hour given in Example 2.4 and so is satisfactory. It must be noted that since the usage rate of air has fallen, so has the charging time.

The maximum number of starts per hour is dependent upon the control mechanism used, and for some equipment a greater number of starts is allowable. To reduce the number of starts per hour there are three alternatives:

1. Alter the compressor capacity, not to be recommended in this case.
2. Alter the size of the receiver, again not the best solution in this case.
3. Adjust the pressures at which the compressor comes on and off load.

The first two options are very expensive and would normally only be considered if there were other more important reasons for their modification.

Adjusting the pressure settings of on-load/off-load values will be investigated. The charge/discharge cycle times of the receiver has to remain greater than 3 min.

Let the difference in pressure settings be P bar, then the volume of air available from the receiver for this pressure difference is:

$$\text{Receiver volume} \times \text{Pressure differences}$$

Therefore,

$$\text{Volume of air available} = 8 \times P \text{ m}^3 \text{ f.a.d.}$$

This volume of air will run the plant for a time given by:

$$\frac{8 \times P}{15} \text{ min}$$

This is the discharge time for the receiver.

To calculate the charge time for the receiver, divide the volume of air which has to flow into the receiver ($8 \times P$ m³ f.a.d. in this case) by the excess flow supplied by the compressor. This is the compressor delivery less the system demand ($35 - 15$ m³/min in this case).

$$\text{Receiver charge time} = \frac{8P}{35 - 15} \text{ min}$$

$$\text{Receiver cycle time} = \frac{8P}{15} + \frac{8P}{20} \text{ min}$$

To give less than 20 starts per hour the cycle time must be greater than 3 minutes. Therefore

$$\frac{8P}{15} + \frac{8P}{20} \geqslant 3$$

$$0.5333P + 0.4P \geqslant 3$$

$$0.9333P \geqslant 3$$

$$P \geqslant 3$$

An increase in air demand will also cause problems with the number of starts per hour.

Example 2.7

Consider the compressor plant designed in Example 2.4 but with the system demand increased from 20 to 25 m³/min f.a.d. Calculate the number of times the compressor switches on load per hour. If the number of compressor starts is to be 20 per hour, determine the receiver pressure at which the compressor switches on load. The off-load pressure is to remain at 7 bar.

Solution

As before, the volume of air stored in the receiver between 7 and 5 bar gauge is 26 m³ f.a.d. At a demand of 25 m³ f.a.d.:

$$\text{Receiver discharge time} = \frac{26}{25}$$

$$= 1.04 \text{ min}$$

Again, with the demand at 25 m³/min f.a.d., there will be $(35 - 25)$ m³/min f.a.d. to charge the receiver. Thus

$$\text{Receiver charge time} = \frac{26}{10}$$

$$= 2.6 \text{ min}$$

$$\text{Receiver cycle time} = 1.04 + 2.6$$
$$= 3.64 \text{ min}$$

This gives 16.5 starts per minute.

Let the difference in the pressure settings be P bar, then the volume of air available from the receiver for the pressure difference is the receiver volume times the pressure difference. Therefore,

$$\text{Volume of air available} = 13 \times P \text{ m}^3 \text{ f.a.d.}$$

This volume of air will run the plant for a time equal to

$$\frac{13 \times P}{25} \text{ min}$$

This is the discharge time for the receiver. To calculate the charge time for the receiver, divide the volume of air that has to flow into the receiver ($13 \times P$ m^3 f.a.d. in this case) by the excess flow supplied by the compressor, which is the compressor delivery less the system demand ($35 - 25$ m^3/min f.a.d. in this case).

$$\text{Receiver charge time} = \frac{13P}{35 - 25} \text{ min}$$

$$\text{Receiver cycle time} = \frac{13P}{25} + \frac{8P}{10} \text{ min}$$

To give less than 20 starts per hour the cycle time must be greater than 3 minutes. Therefore,

$$\frac{13P}{25} + \frac{13P}{10} \geqslant 3$$

$$0.52P + 1.3P \geqslant 3$$

$$1.82P \geqslant 3$$

$$P \geqslant 1.65 \text{ bar}$$

So the on-load pressure is 5.35 bar.

Thus, to obtain a satisfactory number of starts per hour either the maximum or minimum pressure setting has to be altered. Note that the maximum pressure must not exceed the maximum safe working pressure of the system, and the minimum pressure has to be sufficient to give satisfactory operation of the system. Should it not be possible to adjust the pressure difference sufficiently the other alternatives must be considered. For a reduced compressor output or a reduced receiver volume, the simplest solution is to reduce the receiver volume by placing an inert material inside the receiver. If this is not satisfactory, the previously mentioned alternatives apply.

In a multi-compressor installation which uses pressure switches to control individual compressors, there must be sufficient difference in the setting of the switches bringing the compressors on and off load to ensure that hunting does not occur.

2.6.3 In-line receivers

In ring main systems the machines furthest from the compressor plant may suffer from air starvation as the machines nearest the compressor get a preferential supply. On very large installations where the ring main may be 1 km or more in length, it is often practicable to have more than one compressor house and so supply the ring main from several points.

In smaller installations where the fluctuations in supply pressure can cause a machine to malfunction, an air receiver can be installed in the line feeding the machine. The size of the receiver will depend upon the machine demand, the pressure variations allowable and if the receiver is to supply the machine under emergency conditions should the main compressor fail. The receiver will not make up for a shortfall in the air supply, it will only reduce pressure variations.

To size the in-line receiver determine the demand of the machine during the duty cycle time and the variations in the ring main pressure at the take-off point during the period. From these values it is possible to estimate a receiver size to limit the fluctuations in pressure to an acceptable value. A check valve must be installed upstream of the receiver to prevent the air discharging from the receiver into the ring main.

Example 2.8

A machine has an air demand of 0.1 m³ f.a.d. per cycle at a minimum pressure of 4 bar gauge and operates at 25 cycles per minute. The machine is supplied from a ring main in which the pressure varies between 7 and 3 bar gauge linearly over a 20 second cycle time. Estimate the size of a receiver to be placed between the machine and the ring main to maintain a minimum pressure of 4 bar gauge at the machine. Variations in ring main pressure is assumed cyclic, as shown in Fig. 2.18.

The ring main falls below 4 bar pressure for a total time of 5 seconds per cycle variation. This is the time for which the receiver must store sufficient air to operate the machine between a maximum pressure of 5 bar, the average ring main pressure and a minimum of 4 bar.

Solution

Let the receiver volume be V. The volume of air in the receiver at 5 bar is:

$$V \times \left(\frac{5 + 1}{1} \right) \text{ m}^3 \text{ f.a.d.}$$

The volume of air in the receiver at 4 bar is:

$$V \times \left(\frac{4 + 1}{1} \right) \text{ m}^3 \text{ f.a.d.}$$

Therefore, the volume of air available from the receiver is:

$$V \times \left(\frac{5 + 1}{1} \right) - V \times \left(\frac{4 + 1}{1} \right) = V \text{ m}^3 \text{ f.a.d.} \tag{1}$$

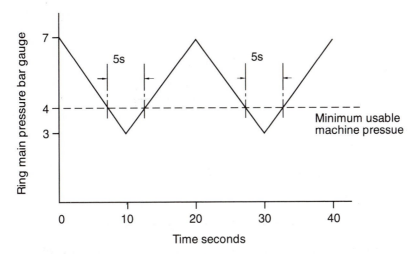

FIGURE 2.18 Cyclic pattern of pressure fluctuations in a ring main system.

The volume of air required by the machine during 5 seconds is the air consumption per cycle times the machine cycles in 5 seconds, that is:

$$\frac{0.1 \times 25 \times 5}{60} = 0.208 \text{ m}^3 \text{ f.a.d.} \tag{2}$$

Thus the minimum volume of the receiver is obtained by equating (1) and (2). That is,

$$V = 0.208 \text{ m}^3$$

The receiver has to be arranged in such a way as to supply the machine with air when the ring main pressure falls below 4 bar gauge. One method is shown in Fig. 2.19.

A large-diameter pipe connects the receiver to the ring main; a smaller diameter take-off pipe supplies the machine. The spring-loaded check valve will close when the

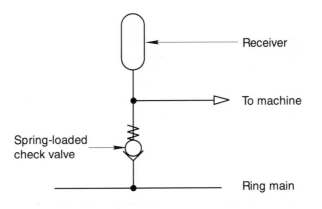

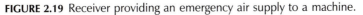

FIGURE 2.19 Receiver providing an emergency air supply to a machine.

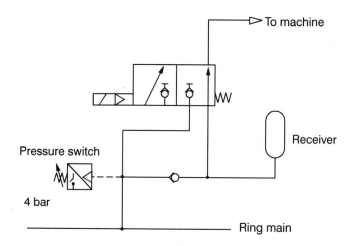

FIGURE 2.20 Pressure sensed switching of an emergency machine air supply.

pressure in the ring main is less than the pressure required at the machine plus the spring setting. This circuit is liable to give problems in operation. As an alternative circuit the pressure in the ring main can be sensed by a pressure switch. When it is above a certain value the machine is supplied from the ring main; if, however, the supply falls below the pressure switch setting, the receiver supplies the machine. A circuit to give this is shown in Fig. 2.20.

Using the circuit shown in Fig. 2.20, as soon as the pressure in the ring main exceeds 4 bar, a pressure switch operates connecting the ring main to the machine, at the same time charging the receiver through the check valve. The setting of the pressure switch can be adjusted if operating conditions vary.

2.6.4 Air leaks and reducing air losses

In large compressed air installations it is not uncommon for the leakage air to account for over 25 per cent of the compressor output. Table 2.2 gives the air leakage rates through various orifices.

TABLE 2.2 Leakage rates through orifices.

Hole diameter (mm)	Air leakage rate	
	l/s	cfm
0.5	0.3	0.6
1.0	1.25	2.6
1.5	3.0	6.3
2.0	5.3	11.2
3.0	11.0	23.3

Orifice leakage rates at 7 bar gauge pressure.

Air leakage is usually through badly fitting joints, damaged pipework, leaky shut-off valves, drains on air filters, badly assembled valve stacks and worn components. The amount of this leakage is often ignored or very much underestimated.

A method of accurately estimating the quantity of leakage air is as follows. Choose a day on which the plant is totally shut down for the test. Make sure that no pneumatic machines are operating and that air is not being used for any purpose. Measure the main air receiver in order to calculate its cubic capacity if this is not already known. Fit an accurate pressure gauge to the receiver. Isolate the receiver from the ring main. Charge the receiver up to its normal working pressure. Note the pressure in the receiver, and if this does not show any change over 15 minutes or so the receiver and valves are relatively air tight; should a leak be detected, this should be cured before proceeding with the test.

The next step is to charge the receiver up to maximum working pressure with the inlet and outlet valves open. This also ensures that the system is charged to full system pressure. Shut off the inlet valve and measure the time for the receiver pressure to fall by a known amount either down to minimum working pressure or by 1 bar. Estimate the volume of the ring main and add this to the volume of the receiver. The time taken for the pressure to fall a known amount has already been measured, so the air leakage rate can be calculated.

Example 2.9

In a test on a ring main it was found that when the receiver was fully charged to 7 bar and then isolated from the compressor it took 10 minutes for the pressure to fall to 6.5 bar gauge. The receiver has a capacity of 20 m³ and the ring main a bore of 100 mm and a length of 800 m. Estimate the leakage rate from the system. If the total input power is 100 kW, estimate the percentage of the input power used in supplying air to the leaks.

Solution

$$\text{Volume of ring main} = \pi \times \left(\frac{0.1^2}{4}\right) \times 800$$

$$= 6.28 \text{ m}^3$$

The total capacity of the receiver and the ring main is

$$20 + 6.28 = 26.28 \text{ m}^3$$

The volume of air released from the receiver and the ring main when the pressure falls from 7 bar to 6.5 bar gauge is

$$(7 - 6.5) \times 26.28 = 13.14 \text{ m}^3$$

This leaks from the system in 600 seconds, so the average leakage rate is

$$\frac{13.14}{600} = 0.0219 \text{ m}^3/\text{s f.a.d.}$$

Or the leakage rate from the system at an average pressure of 6.75 bar gauge is 1.314 m³/min f.a.d.

As previously stated, 75 litre/min f.a.d. at a pressure of 7 bar gauge requires an input of 1 kW. Therefore, the power required to supply 1.314 m³/min at 6.75 bar gauge is:

$$\left(\frac{1.314 \times 1000}{75}\right) \times \left(\frac{6.75 + 1}{7 + 1}\right) = 16.97 \text{ kW}$$

Thus, the percentage of input power used in supplying air leakage is:

$$\frac{16.97}{100} \times 100 = 16.97 \text{ per cent}$$

CHAPTER 3

Valves

Valves fall into three basic categories, each controlling a different facet of the compressed air supply. The valves will control:

- direction
- pressure
- flow.

3.1 Direction control valves

A direction control valve is a device that will, upon the receipt of an external signal, change the direction of, stop or start the flow of air in a particular part of a circuit. Directional valves are usually classified according to the following:

1. The number of connection ports in the body, e.g. 2, 3, 4 or 5.
2. The number of position which the spool can be located.
3. The method of valve actuation.
4. The connection port sizes which can relate to the air flow through the valve.
5. The nature of the internal valve mechanism.

Types of direction control valves

Valve elements   Sliding spool, poppet/diaphragm, rotary spool, rotary disc (plate), slide.
Methods of control   Manual, mechanical, pneumatic pilot signal, hydraulic pilot signal, direct solenoid, solenoid pilot.

3.1.1 Sliding spool valves

Sliding spool valves are probably the most common valves used in the transmission of pneumatic power to the actuator. They come in five basic forms and in sizes ranging from $G.\frac{1}{8}''$ to $G.2''$ portings.

1. Valves with 'O' rings fitted into the bore (Fig. 3.1).
2. Valves with 'O' rings fitted to the spool (Fig. 3.2).
3. Valves with seals bonded to the spool (Fig. 3.3).
4. Valves with 'U' or cup-type seals (Fig. 3.4).
5. Valves with a lapped spool and bore (Fig. 3.5).

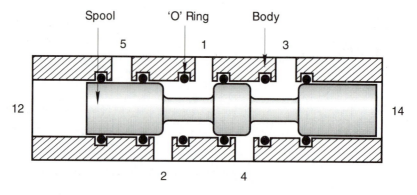

FIGURE 3.1 Typical contoured spool.

Spool valves containing dynamic elastomeric seals offer extremely simple servicing. All that is generally required is the careful removal of the elastomeric seals, examination of the components for mechanical damage and the careful fitting of new seals.

The main advantages of spool valves with seals are:

1. Simple maintenace.
2. Fully balanced spool design allowing air to be connected to any port without creating spool movement that will affect air flow.
3. Relatively simple to attach controls.
4. Stroke limiters can be used to create throttling.

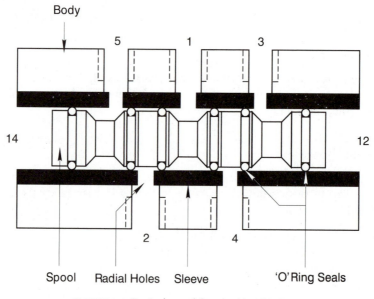

FIGURE 3.2 Typical spool fitted with 'O' rings.

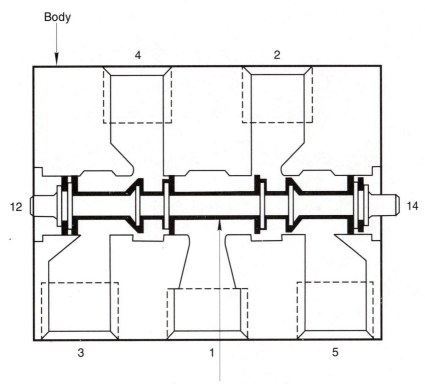

FIGURE 3.3 Typical spool with bonded seals.

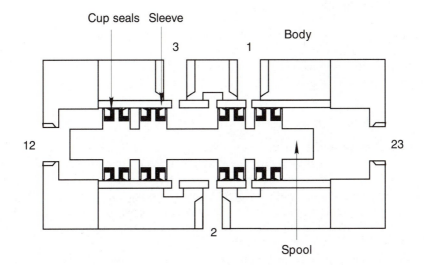

FIGURE 3.4 Typical spool with cup-type seals.

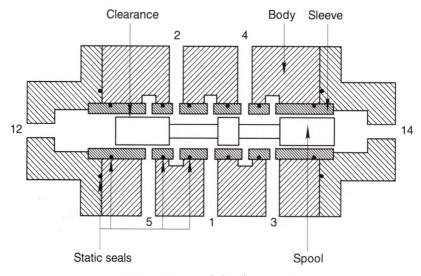

FIGURE 3.5 Typical glandless spool.

5. Three-position conditions can be incorporated having either closed or open centre conditions.
6. Available in forms suitable for mounting on an ISO sub-base.

The main disadvantages of spool valves with seals are:

1. A larger body size is required for an equivalent flow compared to a poppet valve.
2. Higher wear rates than poppet valves.
3. Require lubrication.
4. Continuous leakage.
5. Not suited to high-pressure applications.
6. Require higher operating forces than poppet valves.
7. Slower response time than poppet valves.
8. Require a better quality air than poppet valves.

3.1.2 Glandless spool valves

This type of valve usually incorporates a lapped stainless steel spool and sleeve as a matched pair inside an alloy body. The sleeve is normally supported in the body by a series of 'O' ring seals, this reduces the risk of sleeve distortion due to temperature changes, or the effect of uneven tightening torques on the mounting bolts and pipe fittings.

The spool and sleeve are lapped to give clearance of less than 2 μm; this clearance allows a continuous leakage of air which acts as an air bearing, supporting the spool and making fast, easy spool movement possible.

The maintenance of lapped spool and sleeve valves is limited to the removal of the spool and sleeve, cleaning and the replacement of static seals. Should there be any physical damage to the spool or sleeve, then the pair should be replaced by a new matched set.

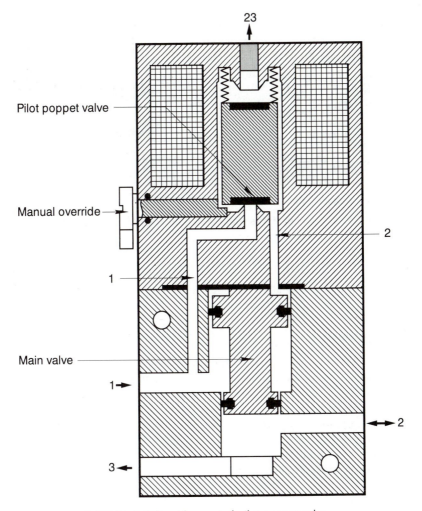

FIGURE 3.6 Solenoid-operated pilot poppet valve.

The main advantages of lapped spool and sleeve valves are:

1. Very long life expectancy (in excess of 100 million cycles).
2. Rapid response.
3. Low actuating forces required.
4. Suitable for higher working pressures than the sliding spool valves with elastomeric seals.
5. Suitable for lubricated or non-lubricated air.

The main disadvantages of lapped spool and sleeve valves are:

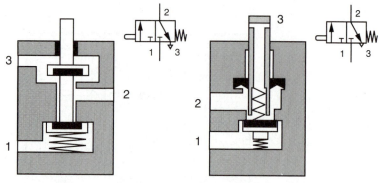

Open transition condition   Closed transition condition

FIGURE 3.7 Poppet valves with open and closed transition conditions.

1. Very expensive.
2. Continual leakage.
3. Require high quality air filtration.

3.1.3 Poppet valves

Poppet valves come in a wide variety of forms and are by far the most prolific valve in pneumatic service, many being used as the pilot section of a solenoid-controlled valve (Fig. 3.6). Poppet valve construction varies in accordance with the valve function and flow requirements, some of which are shown in Fig. 3.7.

The main advantages of poppet valves are:

1. Can operate with lubricant free air.
2. Can operate with inferior quality air.
3. Virtually leak free.
4. Very low wear rates.
5. High flow rates from a relatively small valve.
6. Rapid response.

The main disadvantages of poppet valves are:

1. Cannot easily be serviced.
2. Not suited to reverse porting.
3. Relatively high operating forces required.
4. Possible air loss during changeover.

3.1.4 Rotary spool valves

These are generally small, manually operated machine-mounted valves used to provide a pilot signal to the main power valve. Almost all three-position valves have a closed centre condition. The valves offer simple, stable and easy operation, but are susceptible to contamination and can provide a high resistance to flow (Fig. 3.8).

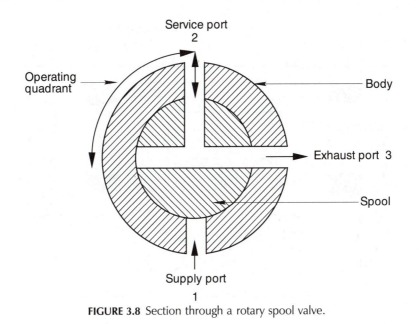

FIGURE 3.8 Section through a rotary spool valve.

3.1.5 Rotary disc (plate) valves

In a rotary disc valve the seal between the mating surfaces is produced by lapping the two halves of the valve. Again, as with rotary spool valves, these tend to be manually operated, machine-mounted valves, having a detented condition with either two or three operating conditions. A small degree of flow control can be achieved by part porting the valve (Fig. 3.9).

3.1.6 Slide valves

This is probably the oldest design of valve used (Fig. 3.10) and is derived from steam

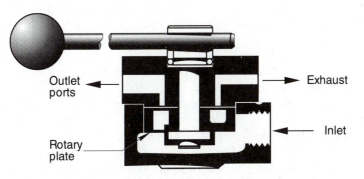

FIGURE 3.9 Section through a rotary plate valve.

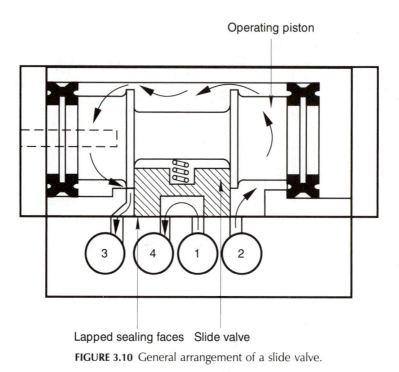

FIGURE 3.10 General arrangement of a slide valve.

engine applications. Its principal use is for the direct operation of double-acting cylinders. As with the two previous valves, a seal is produced by lapping the mating surfaces.

3.1.7 Methods of valve actuation

Manual Lever, push/pull, foot pedal, key, or toggle switch (see Fig. 3.11).
Mechanical Plunger, roller lever, whisker, one-way roller lever, or springs (see Fig. 3.12).
Others Diaphragm, direct pneumatic, direct hydraulic, direct electrical, or electro-
  pneumatic (see Fig. 3.13).

As the majority of manually operated valves offer direct operation of the spool, it is therefore important to avoid excessive loading of the control elements to minimise wear on linkages and push pins. This wear can lead to part porting of the valve with a resulting reduction in system performance.

Where foot pedal-operated valves in particular are used, protective covers should be fitted to prevent accidental operation.

Although many mechanically operated valves use direct actuation, valves that require very light operating forces, e.g. whisker sensors, employ force amplification through the use of an air pilot to operate the main valve spool (Fig. 3.14).

Where direct pneumatic actuation is required, it can be produced by providing a mains air signal directly onto the spool or poppet, e.g. via a position-sensing valve as shown in Fig. 3.15.

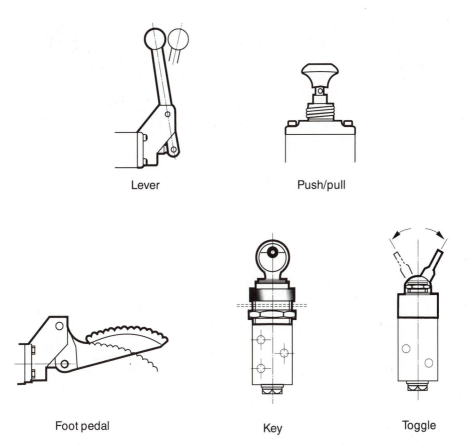

Lever

Push/pull

Foot pedal

Key

Toggle

FIGURE 3.11 A selection of manual valve actuators.

If only low pressure or vacuum signals are available, then the valve may still be operated directly by the signal using a diaphragm to amplify the signal (Fig. 3.16). Minimum operating pressures generally range between 0.0025 and 0.3 bar, with the intermediate range being the most popular.

The direct application of solenoid operation is in the main restricted to the control of pilot air supply to a power valve which can either be attached to the solenoid section or be remotely positioned.

The solenoid is an electro-magnetic device that can be powered with either an a.c. or d.c. electrical voltage. However the majority of applications utilise d.c. solenoids for two reasons.

1. Their ability to interface directly with programmable controllers.
2. The fact that d.c. solenoids do not burn out as quickly as a.c. solenoids should the solenoid not close completely.

Plunger

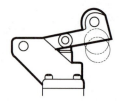

Roller lever

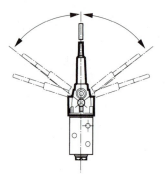

Whisker antenna

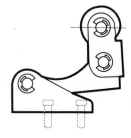

One-way roller lever

FIGURE 3.12 A selection of mechanical valve actuators.

Power supplied to the solenoid usually incorporates a solenoid plug to a respective ISO standard, as shown in Fig. 3.17.

In order to facilitate fault finding, LEDs and manual over-rides are often included in the construction of the valve and its associated electrical actuator.

3.2 Valve specification

Where a valve is to be selected for a specific duty, the specification of that valve should consider the following:

1. Maximum operating pressure.
2. Maximum flow rate required.
3. Maximum/minimum operating temperature.
4. Rate of cycling.
5. Response time.

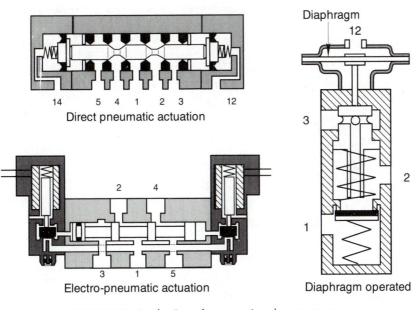

Direct pneumatic actuation

Diaphragm

Electro-pneumatic actuation

Diaphragm operated

FIGURE 3.13 A selection of pneumatic valve actuators.

6. Method of operation.
7. Special material requirements.
8. Air quality.
9. Porting body/manifold.
10. Method of mounting.

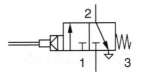

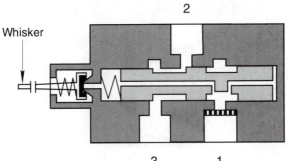

Whisker

FIGURE 3.14 Whisker-actuated valve operation.
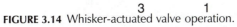

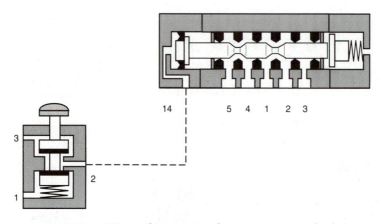

FIGURE 3.15 Direct pilot operation from a remote signal valve.

Although port sizes can give some indication of flow capability, they are by no means good indicators of valve flow performance. To determine more accurately the performance of a valve one must consider its operation across a wide range of conditions.

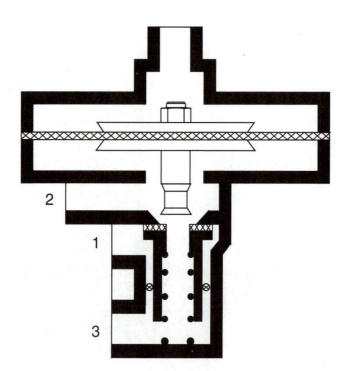

FIGURE 3.16 Amplified pilot operation through a diaphragm.

Rectangular ISO 6952 (DIN 43650 - B)

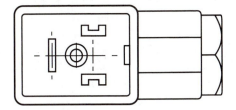

Square ISO 440 (DIN 43650 - A)

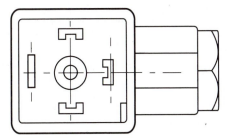

FIGURE 3.17 Solenoid plugs to ISO standards.

Table 3.1 relates relative sizes and approximate average flows.

TABLE 3.1 Approximate air flow rates through a range of valves.

BSP	Metric	Normal bore (mm)	Flow (l/min)
	M5	2–2.7	60–185
$G\frac{1}{8}''$	M10 × 1	3–4	140–400
$G\frac{3}{8}''$	M18 × 1.5	8–9	3800–2000
$G\frac{1}{2}''$	M22 × 1.5	12–13	1800–4500
$G\frac{3}{4}''$	M26 × 1.5	19–20	4000–9000

3.3 Valve performance

When comparing valve performance there can be a maze of flow values given by different manufacturers for different test conditions. In order to produce some standardisation and comparability, various organisations have introduced test standards.

Nominal flow rates
The nominal flow rate (Q_n) for a valve is the flow rate in dm^3/s or l/min across the valve for given entry and exit conditions. Figure 3.18 shows a chart for a particular valve,

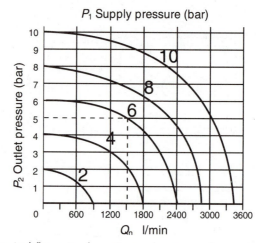

FIGURE 3.18 Nominal flow rates for a given valve over a given pressure drop range.

relating inlet pressure (represented by the curved lines), outlet pressure (vertical axis) and nominal flow rate, Q_n (horizontal axis).

Example 3.1

From the chart in Fig. 3.18, a supply pressure of 6 bar (P_1) and an outlet pressure of 5 bar (P_2) gives a flow rate Q_n of 1500 l/min.

Example 3.2

If, from the previous conditions, the outlet pressure (P_2) falls to 4 bar the flow rate Q_n will increase to 1900 l/min.

Flow factors C_v and K_v

The Fluid Controls Institute of the USA published a standard entitled *Recommended Voluntary Standards for Measurement Procedure for Determining Control Valve Flow Capacity*.

To determine imperial flow rates at 55 °F:

$$Q = C_v \sqrt{[\Delta P((P_s + 14.7) - \Delta P)]}$$

where Q = air flow (ft³/min f.a.d.)
C_v = flow factor for valve
ΔP = pressure drop across the valve (lb/in²)
P_s = supply pressure at the valve inlet (lb/in²).

The above formula can be modified to suit ISO units as follows:

$$Q = 6.844 C_v \sqrt{[\Delta P((P_s + 1) - \Delta P)]}$$

where Q = air flow (litre/s of free air at 12.8 °C)
       C_v = flow factor for valve
       P = pressure drop across the valve (bar)
       P_s = supply pressure at the valve inlet (bar).

A metric version of the C_v flow factor is the K_v flow factor, which is defined as:

The flow of water in litres per minute with a pressure drop of 1 bar across the valve.

$$K_v = 14.28 C_v$$

The CETOP RP50P standard considers the assessment of valve performances in a purely pneumatic form.

Conductance C

The conductance C of a component is the ratio between the flow rate Q across the valve and the inlet pressure P_1 when the air velocity has reached sonic flow conditions at $+20$ °C

$$C = \frac{Q}{P_1}$$

Sonic or choked flow occurs when the inlet pressure P_1 is so high in relation to the outlet pressure P_2 that the air flow is proportional to the inlet pressure P_1 and is independent of the outlet pressure P_2.

Critical pressure ratio b

The critical pressure ratio b is the ratio between the outlet pressure P_2 and the inlet pressure P_1 at which the air velocity achieves sonic speed or (choked flow).

$$b = \frac{P_2}{P_1}$$

With most industrial valves the factor b is usually between 0.25 and 0.5. Theoretically, in an ideal orifice having no flow restrictions, $b = 0.528$. The use of both C and b factors allows the flow performances of different valves to be compared across the whole operational range of valves.

3.4 Valve mounting

Valves are manufactured in two basic forms:

1. Body ported
2. Manifold mounted

3.4.1 Body porting

With body-ported valves metric or BSP parallel threads are by far the most popular; however, with miniature valves M5 threads are becoming the most common. One of the latest innovations is to have the body-ported valve supplied complete with push-in fittings.

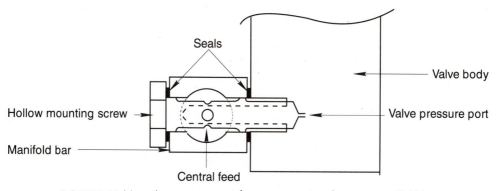

FIGURE 3.19 Mounting arrangement for a pneumatic valve on a manifold bar.

When installing body-ported valves it is important to ensure that the mounting surface is flat, thereby avoiding the risk of distortion to the valve body. For the same reason care must be taken when installing fittings to avoid overtightening.

In order to make the mounting of body-ported valves more cost effective, compact mounting bars can be used. These comprise a bored-out bar with an air supply at one end – or both ends in the case of a large bank of valves – and a number of cross-drillings. The valves are secured to the bar by means of hollow screws fitted into the pressure port of each valve. Leakage is prevented by using 'O' ring seals or soft washers on either side of the bar (Fig. 3.19).

3.4.2 Manifold mounting

One way of reducing the problems caused by fittings on the performance of a valve is to have all the fittings installed in a manifold or base. To this end manufacturers have produced a whole host of sub-bases to suit their own valves. However, among the many sub-bases a range of standard sub-bases has been manufactured to ISO 5599/1. This standard specifies the interface between the valve and the sub-base and covers model sizes 1, 2, 3 and 4 (Fig. 3.20). Care must be taken when replacing ISO interfaced valves and sub-bases since the overall component dimensions can vary between manufacturers, thus causing interchangeability problems between components of different origins.

Sub-base portings can be rear or side entry for use with individual mountings. Where a number of valves are to be accommodated in close proximity, a combination of side and rear entry allows the sub-bases to be fastened together, which reduces the number of connections required. Where large manifolds are constructed, air can be supplied from both ends of the manifold to give an adequate air supply to all valves. Exhaust air can be treated similarly, allowing simple reclassification if required. Many of the non-standard sub-bases and manifolds offer superior flow characteristics and are a much lower priced option.

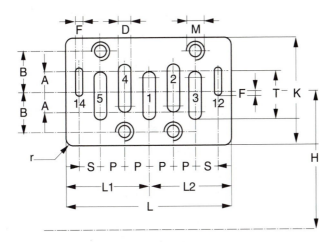

ISO	A	B	D	F	M	T	S
1	9	14	4,5	3	M5	16,5	8,5
2	10	19	7		M6	22	10
3	11,5	24	10	4	M8	29	13
4	14,5	29	13			36,5	15,5

ISO	P	H	r	K	L1	L2	L
1	9	43	2,5	38	32,5		65
2	12	56	3	50	40,5		81
3	16	71	4	64	53		106
4	20	82		74	77,5	64,5	142

FIGURE 3.20 ISO 5599/1 sub-base mounting dimensions for pneumatic directional valves.

3.5 Valve applications

2-port, 2-position valve

These are generally poppet valves, and because of their construction are either normally closed or normally open in their de-energised condition. Typical applications of this type of valve are cylinder hoists, blow guns, hydro-pneumatic control systems and isolating valves. Many manually operated valves require relatively high operating forces and are therefore unsuitable for frequent use. A typical valve is shown in Fig. 3.21.

The use of a tapered stem allows for some control of flow during the initial operation of the valve. This is particularly useful for pneumatic hoist applications.

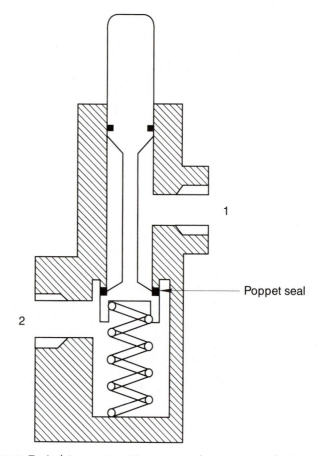

FIGURE 3.21 Typical 2-port, 2-position poppet direction control valve.

3-Port, 2-position valves

These are usually spool or poppet type valves and can be normally open or normally closed in their de-energised condition. Typical applications of this type of valve include the operation of spring-loaded cylinders, and the indirect control of large cylinders by providing pilot control switching for other valves. The pilot valve in most solenoid-pilot direction control valves is in fact a 3/2 poppet valve. Typical 3/2 valves are shown in Fig. 3.22.

4-Port, 2-position/4-port, 3-position valves

These valves tend to be mainly disc type valves suitable for robust manual operation where the inching of a double-acting cylinder may be required. A typical valve application is shown in Fig. 3.23.

5-port, 2-position/5-port, 3-position valves

These are the most commonly used power valves, used for controlling both cylinders and

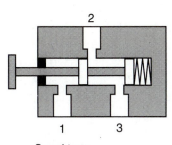

Spool type
normally closed

Poppet type
normally closed

Common symbol

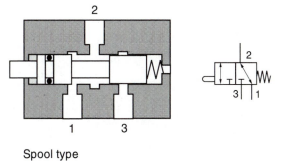

Spool type
normally open

FIGURE 3.22 Typical 3-port, 2-position direction control valves.

motors. They are produced in all forms with an assortment of operational methods. A typical 5-port, 2-position valve is shown in Fig. 3.24.

Where 3-position valves are employed, the centre condition selected is to suit particular operation characteristics of the system.

In the circuit shown in Fig. 3.25 the valve, when in its centre position, causes the cylinder to remain charged with pressurised air in a condition that allows inching in either direction to take place.

In the circuit shown in Fig. 3.26, the valve shown in its centre position causes the cylinder to be pre-exhausted thereby allowing the cylinder to be manually positioned. Note that the load may also move the piston in this condition. When used in conjunction with an air motor this centre condition allows the motor to come to rest gradually, thereby avoiding shock loads which could damage the motor.

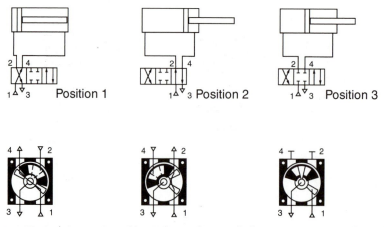

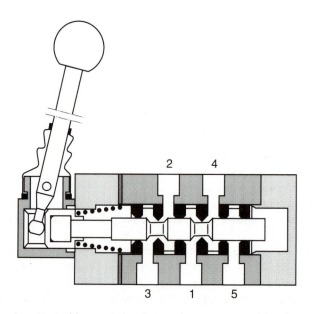

FIGURE 3.23 Typical 4-port, 3-position valve and its symbolic representation with a cylinder.

3.6 Pressure control valves

3.6.1 Pressure relief valves

Within many circuits primary pressure control and protection take place within the air compressor installation boundary. However there are cases where rapid increases in pressure beyond, say, a pressure regulator cannot be controlled by the relief exhaust within that regulator. If this sudden pressure rise is predicted, then fitting an additional

FIGURE 3.24 Typical lever-operated, spring return, 5-port, 2-position direction control valve.

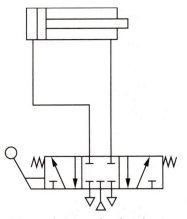

FIGURE 3.25 Symbol diagram of the application of a closed centre, 5-port, 3-position direction control valve.

pressure relief valve between the pressure regulator and the source is essential. Care must be taken when siting additional relief valves to obviate the risk of injury to persons and possible damage to product when the valve discharges. The setting of this secondary relief valve should be 10–15 per cent in excess of the regulated pressure in order to eliminate the risk of premature air leakage.

Direct-loaded relief valves

This poppet-type relief valve can have accuracy limitations up to ± 20 per cent of the relief valve setting. Greater accuracy can be obtained by employing a diaphragm-type relief valve, as shown in Fig. 3.27.

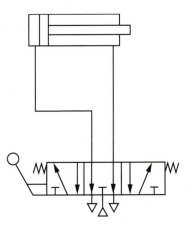

FIGURE 3.26 Symbol diagram of the application of an open centre, 5-port, 3-position direction control valve.

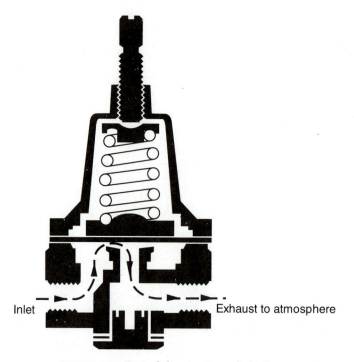

FIGURE 3.27 Typical direct-acting relief valve.

Pilot-controlled relief valve

This relief valve is usually a diaphragm type in which the loading spring has been removed and replaced by air pressure from a controlling pilot signal. The pilot signal is normally set by a remotely situated pressure regulator. It may have a separate air supply to that of the relief valve it is controlling (see Fig. 3.28).

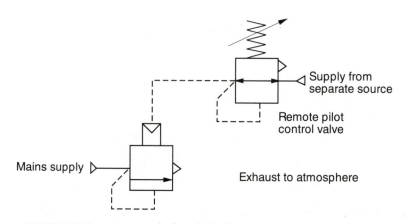

FIGURE 3.28 Remote control of a relief valve using a pressure regulator.

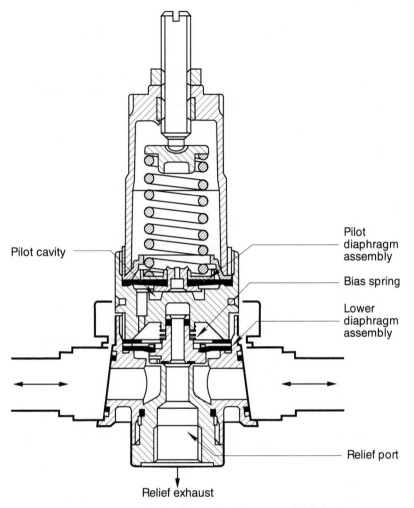

FIGURE 3.29 'In line' integral pilot-operated relief valve.

Pilot-operated relief valve

With this type of relief valve there are two principal sections: the pilot stage and the main stage. Both stages are normally diaphragm-type valves, as shown in Fig. 3.29. The small adjustable pilot spring and pilot diaphragm control the air pressure that sits above the main diaphragm and its light spring. Air is allowed to pass through a small hole in the main diaphragm and onto both the top of the main diaphragm and the underside of the pilot diaphragm. Since the areas on either side of the main diaphragm are equal, the spring biases the main diaphragm into a closed position and, simultaneously, the air pressure acts on the underside of the pilot diaphragm.

When the pressure is sufficient to overcome the pilot spring, the pilot diaphragm lifts and allows the air above the main diaphragm to be exhausted. There is now a pressure

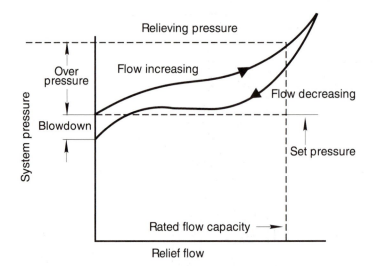

FIGURE 3.30 Typical operating curves for a relief valve.

difference across the main diaphragm which will overcome the light spring and open the main stage of the valve to exhaust. This valve, although slightly slower to react than the direct-acting relief valve, has a much more accurate setting performance.

3.6.2 Pressure terminology

Design pressure
This is the pressure that can be continuously applied to the component/system at the maximum rated design temperature that will permit normal operation of all components without adversely affecting their working life.

Working pressure
This is the pressure at which the system will normally work. The working pressure should be not more than 90% of the design pressure (Fig. 3.30).

Test pressure
This is the pressure at which the system/component is to be tested. It is normally 1.5 times the working pressure which, when applied to the system/component, will not cause leakage or distortion.

Set pressure
This is the pressure at which the relief valve, under normal operating conditions, will open (Fig. 3.30). It is sometimes referred to as the cracking pressure of the relief valve.

Overpressure
This is the pressure increase which is above the set pressure of the relief valve (Fig. 3.30). It is usually expressed as a percentage of the set pressure. The relief valve flow rate which results in an overpressure is generally referred to as the rated flow capacity.

Relieving pressure
Relieving pressure is the set pressure *plus* the overpressure (Fig. 3.30).

Re-seating pressure
This is the pressure at which the valve, after being open, will actually close and prevent relief exhaust (Fig. 3.30).

Proof pressure – type testing
This is the pressure in excess of the test pressure which, if exceeded, may cause leakage or damage to the component. The proof pressure usually varies between 2.5 and 4 times the maximum permissible working pressure.

3.7 Pressure regulators

In order to optimise usage and eliminate excessive forces, pressure regulators are incorporated into pneumatic systems. The most common application is as part of the FRL unit which will control the pressure to a total subsystem. Where a subsystem requires a number of different pressures in different parts of the system then pressure regulators dedicated to the individual parts of the system are installed.

When specifying a pressure regulator there are two basic operating characteristics to be considered:

1. Flow characteristic
2. Regulation characteristic.

The flow characteristic establishes the relationship between the regulated pressure and the rate of flow through the valve. Under ideal conditions there should be no change in regulated pressure regardless of the wide varieties of flow possible. Some typical flow characteristics are shown in Fig. 3.31.

The regulation characteristics establish the relationship between the regulated pressure (downstream) and the primary pressure (upstream). Again, under ideal conditions, there should be no change in regulated pressure even though the primary pressure may fluctuate. A typical regulation characteristic graph is shown in Fig. 3.32.

When selecting a pressure-regulating valve, due consideration should be given to both of the above characteristics.

Although there are many types and sizes of regulators available, their construction falls into two main categories:

1. Diaphragm type
2. Piston type.

Actuation can be 'direct' by some form of adjusting screw or 'pilot-operated' from some remote air pressure source. In general terms diaphragm-type valves tend to be more sensitive to pressure changes, while piston-type valves offer better flow capabilities for their respective sizes. Additionally, the valves can be either 'relieving' or 'non-relieving', the former being by far the more popular. A typical direct-acting, relieving-type diaphragm pressure-reducing valve is shown in Fig. 3.33.

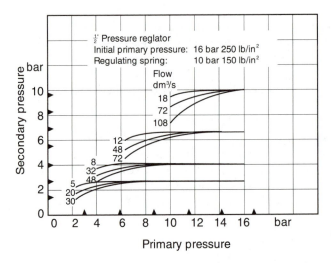

FIGURE 3.31 Typical pressure characteristics for a pressure regulator.

The operation of the valve shown in Fig. 3.33 is as follows:

The working element of this type of pressure regulator comprises a diaphragm which controls a poppet valve via an inter-connecting valve pin and an adjusting spring which is externally loaded by the adjusting screw.

The underside of the diaphragm is connected to the outlet port so that the regulated pressure can act upon the underside of the diaphragm. When the adjusting screw is tightened the main adjusting spring pushes down upon the top of the diaphragm through the valve pin to open the poppet valve and admit pressurised air. As the regulated air pressure increases so does the force on the underside of the diaphragm; this now opposes the adjusting spring causing it to rise until a state of equilibrium exists between the adjusting spring and the downstream regulated air pressure.

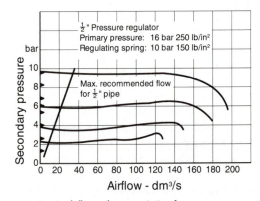

FIGURE 3.32 Typical flow characteristics for a pressure regulator.

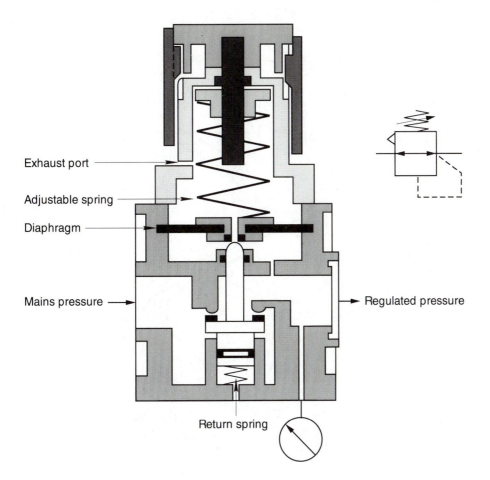

Exhaust port

Adjustable spring

Diaphragm

Mains pressure

Regulated pressure

Return spring

FIGURE 3.33 Typical single-stage pressure regulator.

If there is no flow demand, then when this state of equilibrium occurs the poppet valve will close, thereby preventing further pressurised air from entering the regulated side of the valve. If, however, there is a flow demand then the state of equilibrium will occur with the poppet valve held open sufficient only to allow enough air through the valve to satisfy the demand.

Note. When setting the pressure regulating valve, conditions will alter for 'flow' or 'no flow' conditions.

Should tension now be removed from the adjusting spring by unscrewing the adjusting screw, the regulated pressure and the spring will be unbalanced and the diaphragm will compress the spring and allow regulated air to be discharged through the orifice at its centre. This orifice would normally be closed by the upper end of the valve pin. Under conditions of increased external loading or an increase in downstream temperature, the regulated pressure would rise and be relieved by raising the diaphragm as previously described.

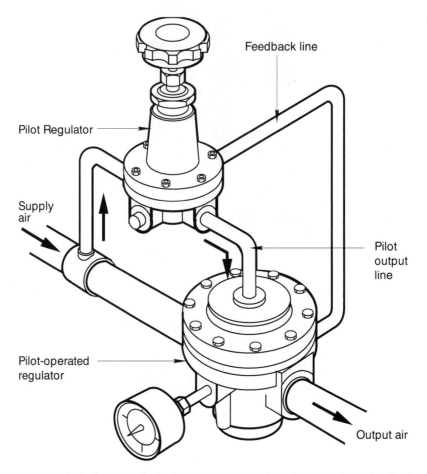

FIGURE 3.34 Typical pilot control of a large pressure regulator incorporating a feedback signal from the main regulator.

Where there is a demand for high flow rates and accurate pressure control, the pilot-operated pressure-reducing valve shown in Fig. 3.34 can be employed.

Remote pilot operation of pressure regulators can also incorporate a feedback line, as shown in Fig. 3.34. With this type of regulator the adjusting spring and screw are removed from above the diaphragm and replaced by a pilot air supply from another pilot pressure-reducing valve. This pilot pressure now provides the force to the top of the diaphragm, thereby opening the poppet valve and allowing air to flow and downstream pressure to rise until a state of equilibrium is obtained across the diaphragm.

The feedback signal is used to enhance the valve's performance under conditions of varying demands and high flow rates.

There are now a number of electrically controlled pressure regulators available which can provide either stepped control of pressure settings or infinitely variable control of pressure settings. The stepped programmable pressure regulator uses a number of individual solenoid-operated valves to pressurise or exhaust their individual piston areas.

The downstream or regulated pressure is thereby a stepped function of the inlet pressure. Since mains air can have pressure fluctuations, the outlet from this valve will also fluctuate in parallel. To avoid this problem the stepped regulator can be supplied from a primary pressure regulator thus stabilising both inlet and outlet pressures.

Multiple pressure selections can be made by interfacing a micro-processor to select the correct pressure for a given application.

The proportional pressure regulator is a variation of the standard manually adjusted diaphragm-type regulator. In this type of regulator the adjusting screw section is replaced by a precision threaded spindle and a DC servo motor drive through a high ratio gearbox. This gives an infinite range of pressure adjustments regardless of fluctuations in supply pressure. By incorporating a downstream pressure transducer, a feedback signal can be provided to create very accurate pressure controls. Some of the latest proportional pressure regulators incorporate proportional force solenoids in lieu of the DC servo motor and screw. The proportional solenoid is controlled by an electronic driver card which may incorporate an electronic feedback for very accurate setting and control.

3.8 Flow control valves

Where it is necessary to control the speed of an actuator this will involve the use of some form of flow regulator. When used in cylinder circuits the flow regulator is usually unidirectional, i.e. flow is controlled in only one direction and free flow occurs in the opposite direction. A typical example is shown in Fig. 3.35.

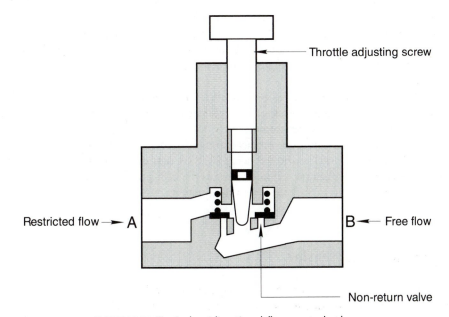

FIGURE 3.35 Typical unidirectional flow control valve.

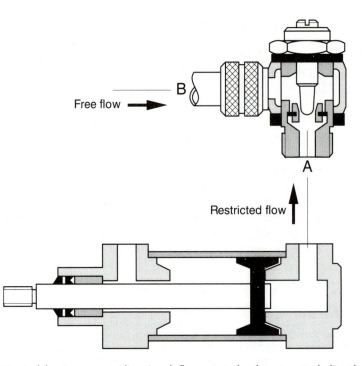

Free flow ➡ B

Restricted flow ↑ A

FIGURE 3.36 Typical banjo type unidirectional flow control valve mounted directly onto a cylinder.

The degree of speed control is dependent upon the shape of the needle. Simple tapers tend to produce a parabolic flow curve, whereas complex tapers tend to give 35 to 45% increased flow per turn of the adjusting screw. Cylinder control is normally achieved by metering the flow of air out of the cylinder, as shown in Fig. 3.36. However, as can be seen from Fig. 3.37, the single-acting cylinder utilises both meter-in and meter-out flow controls.

Meter-in flow controls are also used when the cylinders are so small that there is insufficient air on the annulus side of the cylinder to provide effective control. Unless the flow regulator is severely restrictive it may not appreciatively affect the thrust of the cylinder. Should multiple speeds be required from a cylinder, then an externally controlled flow regulator can be used, as shown in Fig. 3.38. A cam profile would be fitted to the piston rod as shown. When the cylinder extends the cam profile will alter the setting of the flow control valve to match the speed requirements.

3.9 Pneumatic jet sensing

There are many instances in which mechanical or electrical sensing devices are inappropriate. An alternative sensing method is to use an air jet sensor. There are three types of jet-sensing devices available:

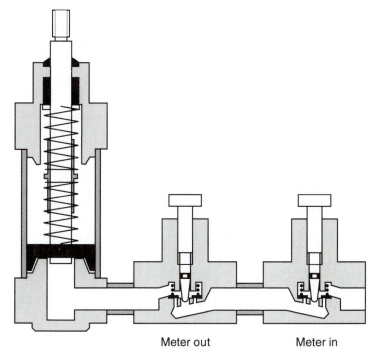

Meter out   Meter in

FIGURE 3.37 Meter-in and meter-out speed control of a single-acting cylinder.

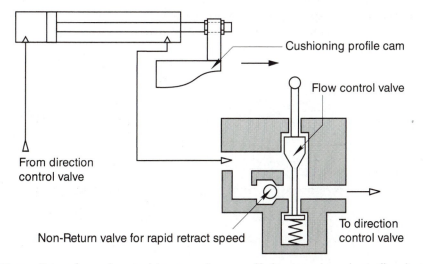

Cushioning profile cam

Flow control valve

From direction
control valve

Non-Return valve for rapid retract speed

To direction
control valve

FIGURE 3.38 External speed control incorporating a profiled cam and mechanically adjusted flow control valve.

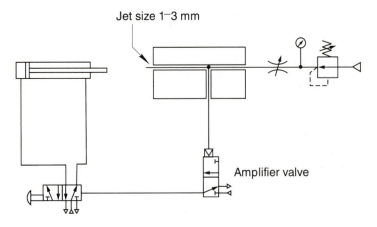

Jet size 1–3 mm

Amplifier valve

FIGURE 3.39 General arrangement of a jet occluded sensing system.

1. Jet occlusion
2. Proximity jet sensing
3. Interruptable jet sensing.

3.9.1. Jet occlusion

This is a relatively simple method, it requires a small hole or jet which is fed by a pressure regulator and a flow control valve. The jet is fitted in lieu of the mechanical part of the trip valve. The supply to the jet sensor is also connected to the pilot of a diaphragm amplifier valve (Fig. 3.39).

The pressure regulator provides an air supply at a pressure slightly above that which is required to operate the diaphragm valve. The flow control valve restricts the air flow to the jet thereby causing the jet's own back pressure to fall below that required to operate

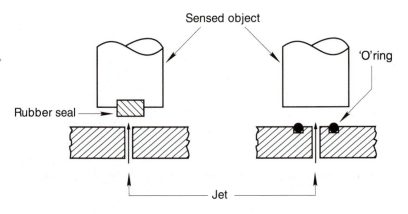

Sensed object

'O'ring

Rubber seal

Jet

FIGURE 3.40 Methods of improving the occlusion of the jet.

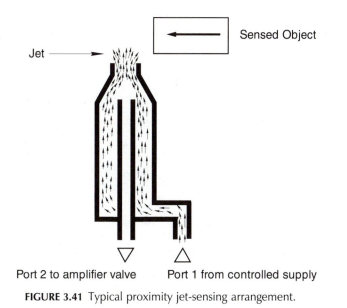

FIGURE 3.41 Typical proximity jet-sensing arrangement.

the amplifier valve. During operation the sensed object approaches the jet sensor and occludes the exhaust path from the jet. The pressure between the jet and the flow control valve now increases and is sensed upon the diaphragm which operates the signal amplifier.

In order to improve the sensor's performance, and hence the diaphragm valve's response, some form of sealing mechanism between the sensor and the sensed object is required (Fig. 3.40). This sealing mechanism can also prevent damage to the jet or sensed object, but will, however, require regular maintenance.

When installing this type of jet sensor one must pay particular attention to the hose diameters and lengths since these can have an adverse effect on valve response times.

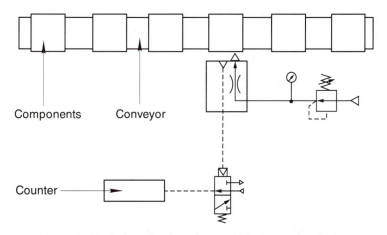

FIGURE 3.42 Typical application of a proximity jet-sensing device.

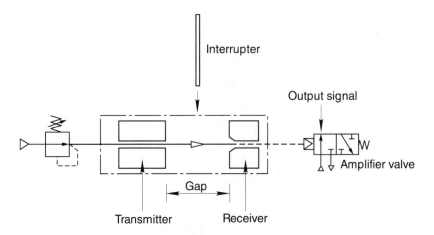

FIGURE 3.43 Typical interruptable jet-sensing arrangement.

3.9.2 Proximity jet sensing

Proximity jet sensors can be used where the sensed object does not come into contact with the structure of the sensor. The jet sensor shown in Fig. 3.41 allows air to be supplied to the jet port 1. While there is no obstruction to the jet, connection 2 is at atmospheric pressure or sometimes has a slight vacuum. When an object approaches the jet sensor

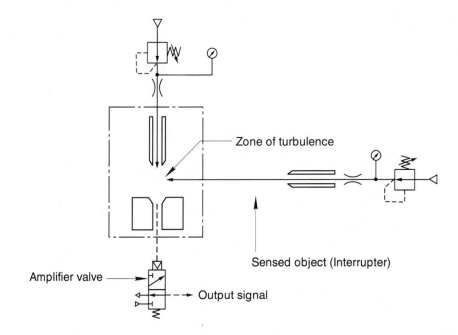

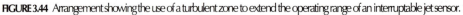

FIGURE 3.44 Arrangement showing the use of a turbulent zone to extend the operating range of an interruptable jet sensor.

the direct release path of the air is obstructed causing the air pressure in the gap between the sensor and the object to increase, and hence provide a positive pressure signal at port 2.

This type of sensor is often used for counting. Figure 3.42 shows a typical application in which the signal pulse is amplified and passed to a pneumatic counter.

Proximity jet sensors are usually only effective up to range of about 2 mm.

3.9.3 Interruptable jet sensing

Where relatively close contact with a sensed object is not acceptable then an interruptable jet sensing device may be suitable. With this type of sensor the gap between the transmitter jet and the receiving jet may be up to 12 mm. A typical arrangement is shown in Fig. 3.43.

In this arrangement the air from the transmitter opposes the air escaping from the receiver, thereby maintaining the signal on the amplifier valve. When the air between the receiver and the transmitter is interrupted, the signal on the diaphragm valve is lost and the valve switches. If a wider gap is required then this sensor is unsuitable in its standard form. In order to accommodate the increased gap requirement, an additional jet can be introduced at right angles to the interrupted jet sensor as shown in Fig. 3.44.

With this arrangement the jet from the intergral transmitter is disturbed by the external jet meeting it at right angles. This causes a turbulent area at the receiver and hence there is an absence of a direct signal and the diaphragm valve is inoperative.

When the sensed object breaks the external jet signal the area of turbulence is eliminated, the internal jet is collected by the receiver and is then applied to the diaphragm, switching the valve and providing an output signal for the next step in the sequence.

This type of sensor can function with gaps up to 300 mm but may require shrouding from external air disturbances.

CHAPTER 4

Actuators

A pneumatic actuator is used to convert the energy of the compressed air into mechanical work. There are three types of pneumatic actuators:

- Linear – a cylinder
- Rotary (continuous) – a motor
- Rotary (limited angle of rotation) – a semi-rotary actuator.

4.1 Pneumatic cylinders

These can be divided into three categories:

- Single rod
- Double or through rod
- Rodless.

All these categories can be single acting (powered by compressed air in one direction and returned by an internal spring or an external force) or double acting (pneumatically powered in both directions).

4.1.1 Single-rod single-acting cylinders

One design of this type of cylinder is shown in Fig. 4.1. The piston is extended by the air supply and retracted by a light spring. It is possible to obtain a single-acting cylinder

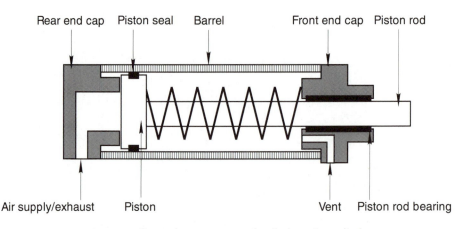

FIGURE 4.1 General arrangement of a single-acting cylinder.

which is retracted by the air supply and extended by an internal spring. Spring-return cylinders are limited in their stroke by the length of the return spring. A double-acting cylinder can be used in the single-acting mode provided an external force is used to return the piston and the return port is left open to act as a breather.

Single-acting cylinders offer three principal advantages over double-acting cylinders.

1. Generally a lower priced installation.
2. Simpler pipework hence more compact.
3. In case of failure of the air supply they will return to their 'at rest' condition.

4.1.2 Single-acting diaphragm cylinders

Diaphragm cylinders, similar to the one shown in Fig. 4.2, are designed principally to exert relatively high forces but over very short strokes. Cylinder sizes range from 8 to 63 mm bore with stroke lengths up to about 10 mm. This type of actuator is often employed for clamping, ejecting and forming. With this design of actuator, friction is negligible, therefore it is well suited to lower pressure applications.

4.1.3 Air bellows

The modern air bellows is a derivative of a vehicle suspension system. It comprises one,

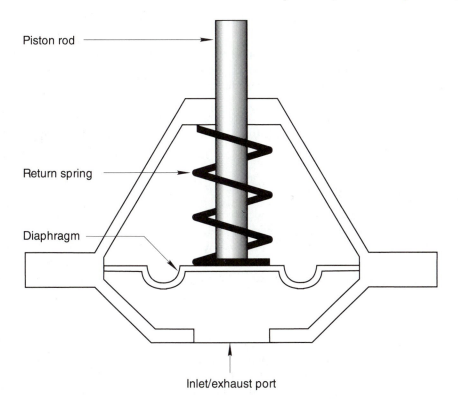

Piston rod

Return spring

Diaphragm

Inlet/exhaust port

FIGURE 4.2 General arrangement of a single-acting diaphragm cylinder.

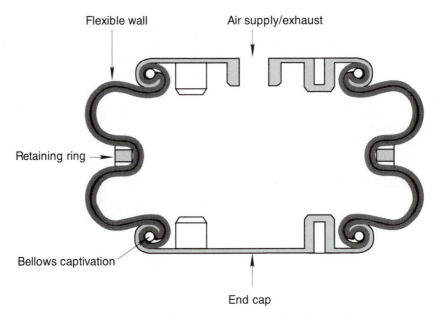

FIGURE 4.3 General arrangement of a single bellows unit.

two or three convolutions of fabric reinforced synthetic rubber (Fig. 4.3) which is almost frictionless in operation.

Owing to the flexible construction of the bellows unit, its mounting is much less rigid that that of a conventional pneumatic cylinder. It is possible to achieve an axial load movement of up to 30 °C or a perpendicular offset of up to 10 mm and still operate successfully (see Fig. 4.4).

Bellows units may be employed where short strokes with high forces are required. Typical applications include lifting platforms, tables, clamping devices, tensioning devices, constant levelling and vibration isolators.

4.1.4 Double-acting cylinders

Where reciprocating action with moderate forces in either direction is required, the application of a double-acting cylinder (Fig. 4.5) is generally suitable. Owing to the difference of area between the full bore side and the annulus side of the piston, the extending force will be greater than the retracting force at the same supply pressure. The retract speed will be faster than the extend speed given the same supply conditions.

4.1.5 Tandem cylinders

The arrangement shown in Fig. 4.6 illustrates how two cylinders can be assembled in a series configuration. This style of cylinder arrangement will produce almost double the thrust of a single cylinder while still retaining the same cross-sectional area. When operating this type of cylinder in the extend mode, compressed air would be supplied to ports 1 and 3 with ports 2 and 4 being exhausted.

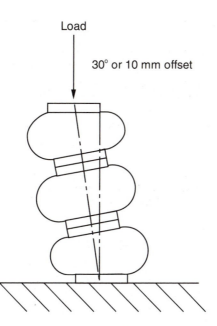

Load

30° or 10 mm offset

FIGURE 4.4 Approximate limitation of offset loading on a multiple bellows unit.

Depending upon the retract force required, compressed air can be supplied to either ports 2 or 4, or to ports 2 and 4. If the retract force is minimal then supplying only one of the ports with compressed air will result in a reduction in the overall usage of compressed air and, in these circumstances, the remaining port must be vented.

4.1.6 Duplex cylinders

The duplex cylinder arrangement can be produced by joining either the two rods of two double-acting cylinders together or by joining their full bore end caps together, as shown in Fig. 4.7.

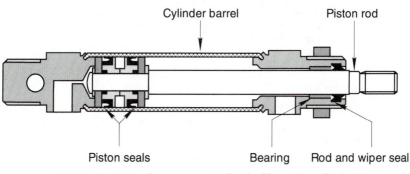

Cylinder barrel   Piston rod

Piston seals   Bearing   Rod and wiper seal

FIGURE 4.5 General arrangement of a double-acting cylinder.

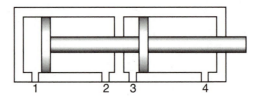

FIGURE 4.6 General arrangement of a tandem cylinder.

If cylinders with the same stroke are selected, then the unit can achieve three distinctive positions. If, however, the two cylinders have different stroke lengths, then four distinct positions will be achieved, as in Fig. 4.8.

4.1.7 Double-rod or through-rod cylinders

This type of cylinder, shown in Fig. 4.9, offers the ability to provide equal thrusts and equal speeds in either direction and the possibility of working from either or both ends.

4.1.8 Impact cylinders

Where stamping, punching, cutting or similar operations are carried out, impact cylinders can be used. A typical impact cylinder is shown in Fig. 4.10. The impact cylinder derives

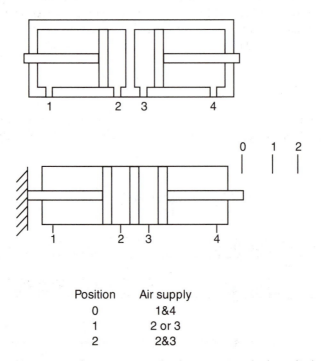

Position	Air supply
0	1&4
1	2 or 3
2	2&3

FIGURE 4.7 General arrangement of a three-position duplex cylinder.

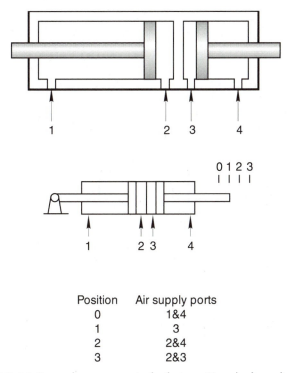

Position	Air supply ports
0	1&4
1	3
2	2&4
3	2&3

FIGURE 4.8 General arrangement of a four-position duplex cylinder.

its power from the expansion of the compressed air accelerating the piston rod and tooling to a high velocity prior to its striking the workpiece. Because of the high energies involved the piston and piston rod are physically much heavier and stronger than that of a standard double-acting cylinder.

Method of operation

Figure 4.11 shows the sequence of events in the operation of an impact cylinder.

Stage 1:   With the direction control valve in the de-energised condition, compressed air is supplied to the annulus side of the piston while the full bore side of the system is exhausted to atmosphere.

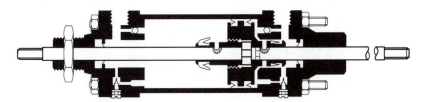

FIGURE 4.9 General arrangement of a through- or double-rodded cylinder.

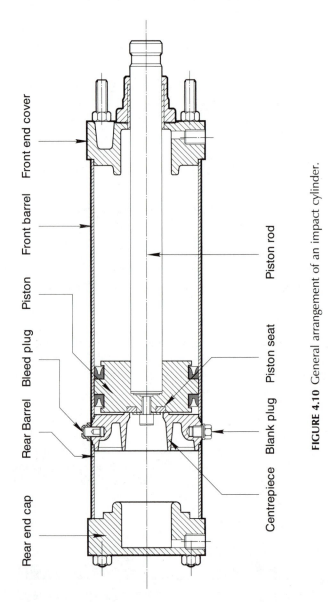

Rear end cap   Rear Barrel   Bleed plug   Piston   Front barrel   Front end cover

Centrepiece   Blank plug   Piston seat   Piston rod

FIGURE 4.10 General arrangement of an impact cylinder.

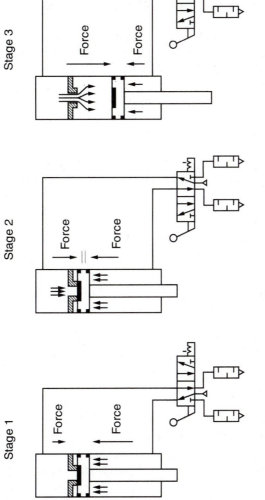

FIGURE 4.11 Sequence of operations of an impact cylinder cycle.

Stage 2:   The direction control valve is operated and the air supply paths are reversed. The annulus side of the piston is exhausted and the compressed air is supplied to the reservoir. Since only about 11 per cent of a full bore side of the piston is exposed to the incoming air supply there is a delay while the annulus side pressure decays and forces on the piston become balanced.

Stage 3:   The continuing changes of pressure on either side of the piston create an out-of-balance force which begins to extend the piston. As soon as the piston seal on the reservoir is broken, the full area of the piston is exposed to the compressed air from the reservoir, resulting in a large increase in force and a rapid acceleration of the piston.

Upon completion of the working stroke the direction control valve is de-energised and the piston retracts.

In the majority of systems with pressures of 5 to 6 bar at the cylinder, maximum energy is attained after about 7 mm of piston travel. Should greater energies be required then higher pressures may be used, but in order to attain maximum energy greater stroke lengths are required (see Table 4.1).

TABLE 4.1 Energy potential for impact cylinders.

Bore size (mm)	Energy	
	Joules (N m)	ft lb
50	25	19
80	70	51
100	128	92
150	255	185

*Energy within impact cylinders is a measure of the ability of the impact blow to do work. The above table of energies assumes an effective pressure at the cylinder of 5.5 bar gauge, a free running cylinder and no additional tooling load.

Installation of impact cylinders

When installing an impact cylinder the plane in which it acts is not critical; however, it is generally more successful to mount the cylinders so that they act either vertically up or vertically down to the point of impact. A simple frame comprising top and bottom plates joined by four tie-rods is shown in Fig. 4.12. The frame should be of sufficient strength, weight and stiffness to resist thrust recoil and impact forces. Owing to the repeated shock loadings, cast steel frames are not generally suitable. Where fabricated frames are used care should be taken with the quality of the welds and the nature of their loads.

Where possible the frame should be secured to a heavy bench or mainframe structure since repeated operations can cause the frame to 'walk'.

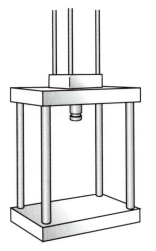

FIGURE 4.12 General arrangement of an impact cylinder support frame.

4.2 Cylinder sizing

When sizing pneumatic cylinders the first step is determining the output requirements, i.e. the nature and magnitude of the load and its speed.

4.2.1 Static force calculation

Figure 4.13 shows a double-acting cylinder, using the formula:

$$\text{Force} = \text{Pressure} \times \text{Area}$$

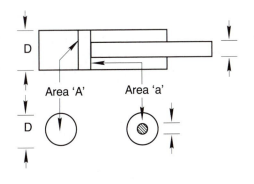

Diameter D = Full bore diameter
Diameter d = Rod diameter
Area A   = Full bore area
Area a   = Annulus area

FIGURE 4.13 Double-acting cylinder.

It is possible to determine the maximum theoretical static thrust/pull.

The theoretical static thrust of a pneumatic cylinder is calculated as follows:

$$F_{ext} = P \times A$$

where F_{ext} = extension force (thrust) (N)

P = static pressure (N/m²)

A = cross-sectional area (full bore side) (m²)

$= \pi D^2/4$

$$F_{ret} = P \times A$$

where F_{ret} = retraction force (pull) (N)

A = cross-sectional area (annulus side) (m²)

$= \pi(D^2 - d^2)/4$

Note: 1 bar = 100 000 N/m²

1 kPa = 1000 N/m².

To determine the static thrust the efficiency of the actual cylinder must be taken into account, which, for operating pressures in the 4 to 7 bar range, is in the order of 94 to 98 per cent. Should operating pressures fall, the cylinder efficiency will also fall. Hence,

$$F_{actual} = P \times A \times \text{Efficiency}$$

In practical design applications the efficiency of the cylinder has little or no effect on the cylinder size.

Example 4.1

A double-acting cylinder is required to clamp a workpiece with an actual force of 3 kN. The minimum supply pressure is 5 bar, cylinder return is at minimal load. Calculate the cylinder size required assuming an efficiency of 96 per cent.

Solution

$$F_{ext} = P \times A \times \text{Efficiency}$$

$$A = \frac{F_{ext}}{P \times \text{Efficiency}}$$

Therefore

$$A = \frac{3000 \text{ N}}{5 \times 10^5 \text{ N/m}^2 \times 0.96}$$

$$= 0.00626 \text{ m}^2$$

Now

$$A = \frac{\pi D^2}{4}$$

Therefore

$$D = \sqrt{(4A/\pi)}$$

$$= \sqrt{\left(\frac{4 \times 0.00626}{\pi}\right)} \text{ m}$$

$$= 0.0893 \text{ m}$$

$$= 89.3 \text{ mm}$$

In this instance one would select a 100 mm bore cylinder since this is the nearest standard bore cylinder that will satisfy the force requirement (see Table 4.2).

4.2.2 Dynamic cylinder forces

When a piston moves within a cylinder there are a number of factors that influence that movement. The energy supply is determined by the pressure and flow rate. The total load is more complex and comprises:

(a) the nature and magnitude of the output
(b) seal friction
(c) acceleration
(d) flow resistances
(e) back pressures.

While the nature and magnitude of the output load is generally easy to determine, the other factors can be difficult to calculate. For this reason many system designers add about 50 per cent to the actual static load to account for the additional resistances, some of which are variable throughout the movement of the cylinder.

Consider the diagram shown in Fig. 4.14, where the relationship between input pressure, load, speed and back pressure is illustrated.

Upon initiation the power valve moves causing the cylinder to start its extend stroke. At this point it can be seen that the high-pressure air in the annulus side of the cylinder begins to decay as it is opened to atmosphere. Simultaneously, the pressure in the full bore side of the cylinder begins to rise towards the minimum system pressure. During this period the force resisting motion is greater than the force to create motion, therefore there is no movement. At the point where this condition is reversed (i.e. the force to create motion is greater than that resisting motion) the piston will 'break out' and commence its acceleration.

As the piston accelerates air is driven out of the annulus side, increasing the back pressure on the piston which gradually reduces the acceleration rate. At the point where the driving and resisting forces are equal, the piston velocity will be a constant.

When the cylinder enters the cushion the exhausting air is restricted, thereby inducing an increased back pressure. This rise in back pressure upsets the balance of

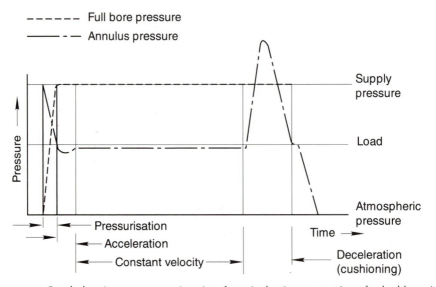

FIGURE 4.14 Graph showing pressure against time for a single piston extension of a double-acting cylinder.

forces and creates piston deceleration. Along with a rapid rise in pressure on the annulus side of the piston there is also a rise in air temperature as a result of (a) degrading the kinetic energy of moving parts and (b) increased air friction. Let us now consider the air leaving the annulus side of the cylinder. The maximum air velocity at the outlet port occurs when the air flow is sonic. Air flow is said to be sonic when its velocity reaches the speed of sound. In theory, when the pressure drop is equal to or greater than $0.53P_1$ sonic flow will occur (Fig. 4.15).

Since pneumatic circuitry can be complicated – involving varying lengths of pipe, fittings and other restrictions – it can be estimated that a pressure drop of 1.4 bar will normally be sufficient to create sonic flow. Where non-cushioned cylinders are used there is little point in generating pressures in the annulus side of the cylinder since this will not affect the exhausting air velocity, but merely induce a higher driving pressure than is absolutely necessary. However, where cushioning is used, an additional back pressure is required to enable the cushion to function adequately. In this case the back pressure is usually in the order of 30 to 40 per cent of the driving pressure.

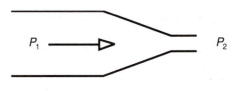

$$P_1 - P_2 = \Delta P = > 0.53\,P_1 = \text{Sonic flow}$$

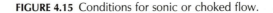

FIGURE 4.15 Conditions for sonic or choked flow.

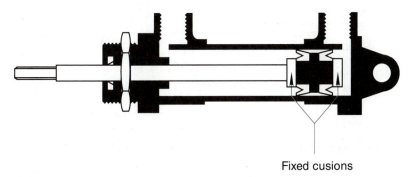

Fixed cusions

FIGURE 4.16 General arrangement of the elastomeric cushioning of a double-acting cylinder.

4.2.3 Cushioning of cylinders

The cushioning of a pneumatic cylinder is required when the kinetic energy of the moving mass is dissipated at the end of its stroke, causing damage to the piston and/or end caps. In cylinders up to 25 mm bore, cushioning is usually achieved by incorporating an elastomeric sleeve or buffer into the end cap, as shown in Fig. 4.16. The degree of cushioning is fixed regardless of load.

For cylinders above 25 mm bore, cushioning is provided which can be adjusted to suit the kinetic energy of the moving parts. A typical example is shown in Fig. 4.17.

As the piston approaches the end of its stroke the tapered cushioning boss enters the main outlet port, creating an initial restriction to the outflowing air. The resulting increase in pressure opposes the piston movement, begins to dissipate the kinetic energy and causes initial retardation. At the point where the cushioning boss fills the outlet bore,

Variable throttle

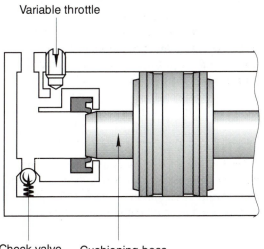

Check valve   Cushioning boss

FIGURE 4.17 General arrangement of the internal cushioning of a cylinder.

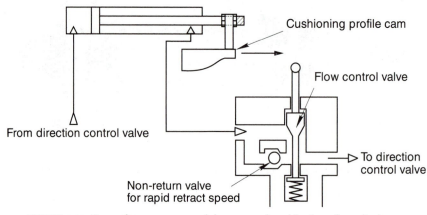

FIGURE 4.18 General arrangement of the external cushioning of a cylinder.

the remaining air has to pass through the throttle; the increased resistance to flow induces an even higher back pressure causing further retardation of the piston and load.

With certain applications when the internal cushioning may be ineffective (e.g. for very heavy loads or exceptionally high velocity pistons), external cushioning has to be employed.

A typical external cushion is shown in Fig. 4.18. As the piston extends, the ramp profile attached to it comes into contact with the adjustable throttle, and the exhausting air is restricted as the valve is gradually closed. The retardation is determined by the ramp profile. When cushioning is employed, a non-return valve is usually incorporated to provide a normal response in the opposite direction.

It is vital to correctly select, install and adjust a cushioned cylinder. If the cylinder selected does not reach the end of its stroke then the cushioning effect will be greatly reduced. Somewhere in the region of 50% loss of cushioning is brought about by the piston having as little as a 5 mm reduction in its potential stroke.

When adjusting cylinder cushions it is important to eliminate as far as possible the metallic 'knock' resulting from the piston striking the end cap too hard. This is achieved by closing the cushioning screw further. Excessive adjustment can cause the piston to bounce and the load to become unstable.

Example 4.2

A pneumatic cylinder is required to move 200 kg packs of paper 600 mm up a 60° incline. The coefficient of friction is 0.15. It is to be assumed that the acceleration of the load will occur within the cushioned length (30 mm) and that the load will attain a velocity of 0.6 m/s. The maximum pressure available at the piston is 5 bar gauge, determine:

(a) the actuator size required
(b) the air consumption if the cylinder operates at 15 cycles/min.

Assume internal frictional resistance plus other losses to be equivalent to 10 per cent of the total force available.

Solution

Determine the total force opposing motion

$$F_{tot} = F_1 + F_f + F_a$$

$$F_1 \text{ (gradient force)} = m \times g \times \sin 60$$

$$= 200 \times 9.81 \times 0.866$$

$$= 1699 \text{ N}$$

$$F_f \text{ (frictional force)} = m \times g \times \cos 60 \times \mu$$

$$= 200 \times 9.81 \times 0.5 \times 0.15$$

$$= 147 \text{ N}$$

$$F_a \text{ (acceleration force)} = \text{mass} \times \text{acceleration}$$

from the equation of motion $V^2 = U^2 + 2as$

if we accelerate from rest $a = \dfrac{V^2}{2s}$

$$\therefore \text{ mass} \times \text{acceleration} = \dfrac{mV^2}{2s}$$

$$= \dfrac{200 \times 0.6^2}{2 \times 0.03}$$

$$= 1200 \text{ N}$$

$$F_{tot} = F_1 + F_f + F_a$$

$$= 1699 + 147 + 1200$$

$$= 3046 \text{ N}$$

Internal friction $= 0.1 \times 3046 \text{ N}$

Total resistance $R_T = 1.1 \times 3046 \text{ N}$

$$\text{Piston diameter} = \sqrt{\left(\dfrac{4 \times R_T}{\pi \times P}\right)}$$

$$= \sqrt{\left(\dfrac{4 \times 3046 \times 1.1}{\pi \times 500\,000}\right)}$$

$$= 0.085 \text{ m}$$

$$= 85 \text{ m}$$

The nearest standard cylinder that would satisfy this application would be 100 mm diameter (see Table 4.2).

To ascertain the quantity flow requirements it is necessary to determine the displacement of the cylinder and the number of cycles per minute.

1. Cylinder (extend)

$$Q_{ext} = \pi \times \left(\frac{D^2}{4}\right) \times L \times n \times \left(\frac{P_1 - P_0}{P_0}\right)$$

2. Cylinder (retract)

$$Q_{ret} = \pi \times \left(\frac{D^2 - d^2}{4}\right) \times L \times n \times \left(\frac{P_1 - P_0}{P_0}\right)$$

3. Cylinder (extend and retract)

$$Q_{tot} = \pi \times \left(\frac{2D^2 - d^2}{4}\right) \times L \times n \times \left(\frac{P_1 - P_0}{P_0}\right)$$

where Q_{ext} = quantity of air to extend the cylinder per minute (m³/min)

Q_{ret} = quantity of air to retract the cylinder per minute (m³/min)

Q_{tot} = quantity of air to extend and retract the cylinder per minute (m³/min)

D = diameter of full bore side (m)

d = diameter of piston rod (m)

L = length of stroke (m)

n = number of cycles per minute

P_1 = supply pressure (bar)

P_0 = atmospheric pressure (bar)

Solution

$$Q = \pi \times \left(\frac{2D^2 - d^2}{4}\right) \times L \times n \times \left(\frac{P_1 - P_0}{P_0}\right)$$

$$= \pi \times \left(\frac{2 \times 0.1^2 - 0.025^2}{4}\right) \times 0.8 \times 15 \times \left(\frac{5 - 1}{1}\right)$$

$$= 0.73 \text{ m}^3/\text{min}$$

$$= 73 \text{ dm}^3/\text{min}$$

$$= 12.17 \text{ dm}^3/\text{s}$$

$$= 12.17 \text{ l/s}$$

To aid the system designer when selecting a cylinder many manufacturers produce charts and tables relating to both air consumption and effective cylinder forces, as shown in Table 4.2.

TABLE 4.2 Theoretical forces and air consumption for a range of cylinder sizes*

Piston dia. (mm)	Rod dia. (mm)	Thrust (N)	Pull (N)	Consumption (dm^3/cycle/mm stroke)
10	4	55	45	0.0011
16	6	140	120	0.003
20	8	220	190	0.0046
25	10	340	290	0.007
32	12	560	480	0.012
40	16	880	740	0.0198
50	20	1 375	1 150	0.031
63	20	2 180	1 960	0.0495
80	25	3 500	3 175	0.076
100	25	5 500	5 150	0.125
125	32	8 600	8 000	0.195
160	40	14 100	13 200	0.32
200	40	22 000	21 100	0.5
250	50	34 350	33 000	0.78
320	63	56 300	54 100	1.29

*The theoretical forces and consumptions in the above table are based upon an air pressure of 7 bar gauge and do not take account of frictional or volumetric losses.

4.2.4 Cylinder speeds

The necessity to determine cylinder performance within a system can be very critical to the production rate of a machine. However, in many instances an effective approximation is sufficiently accurate for all practical purposes, especially if the actual performance is superior to the approximation.

Let us now consider a double-acting, single-rod cushioned cylinder controlled by a symmetrical 5-port, 2-position direction control valve (Fig. 4.19), with the following nomenclature:

D = full bore diameter (mm)
d = piston rod diameter (mm)
L = stroke of the cylinder (mm)
L_c = length of the cushion (mm)
P = supply pressure (bar)
m = mass of the load (kg)
f = constant frictional force (N)
c_e = conductance of the exhaust flow path (l/min)

In order to approximate the total stroke time one must identify the stages of the cycle.

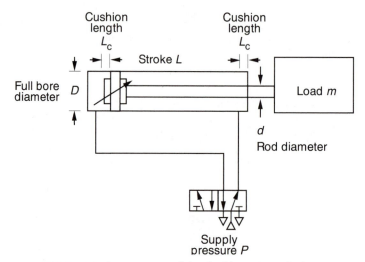

FIGURE 4.19 Arrangement of the double-acting cylinder.

Stage 1 (T_1)

This is the response time for the direction control valve that can be obtained from the valve manufacturer's technical specification, and is usually a constant for that valve.

Stage 2 (T_2)

This is the time taken to reach the cushion stage of the cylinder and includes the time taken to raise sufficient pressure in the cylinder to overcome the frictional resistance, the inertia of the loads and the acceleration forces. The nature of the load will affect the time (T_2), a large frictional force will produce a long dwell period prior to the commencement of motion, and a very short acceleration period. If the load mass is relatively frictionless, there is a very short dwell period and a long acceleration period.

After the dwell and acceleration periods the uniform velocity is strictly dependent upon the supply pressure and load condition. However, for the majority of conditions where the supply pressure exceeds 4 bar and the load is less than 50% of the maximum cylinder thrust at the working pressure, the exhaust flow will be choked. Under these conditions the steady cylinder velocity is governed only by the cross-sectional area of the cylinder and the exhaust path conductance (C_e).

$$\text{Quantity flowing} = \text{Cylinder area} \times \text{Cylinder velocity} \times 10^{-3} \ (\text{l/s})$$

$$= C_e$$

$$V_{\text{max}} = \frac{10^3 \times C_e}{a} \tag{1}$$

where V_{max} = maximum velocity of the cylinder (m/s)
  a = cross-sectional area of the exhaust side of the cylinder (mm²)
  C_e = conductance (l/bar (s))

To determine T_2 it is useful to work on an average velocity V_{ave} of the cylinder:

$$V_{ave} = K \times V \tag{2}$$

where K is a constant. Depending upon the nature of the load, K is a variable theoretically between 0.5 and 1.0. For general purposes a conservative value of $\frac{2}{3}$ is adopted.

The time required to reach the cushion T_2 can be approximated by:

$$T_2 = \frac{\text{Distance travelled}}{\text{Average velocity}}$$

$$= \frac{L - L_c}{\frac{2}{3} \times 10^6 \; C_e/a} \simeq \frac{1.2 \; (L - L_c) \; D^2}{C_e \times 10^6} \tag{3}$$

Although the above will provide a fairly accurate indication of the maximum piston velocity it may not be practicable to operate at these velocities due to the cushion performance, e.g. bounce, excessive cushion pressures or impact forces at the end of cushioning.

Not only the driving force but also the necessary braking force with the cushion must be considered. Assume that the kinetic energy of the mass is to be effectively dissipated throughout the entire cushion length. If we consider an energy limit E on the cushion, we can now determine an alternative V_{max}:

$$E = \tfrac{1}{2} m (V_{max})^2$$

therefore,

$$V_{max} = \sqrt{\frac{2E}{m}} \tag{4}$$

A new T_2 can be calculated

$$T_2 = \frac{(L - L_c) \times m^{0.5}}{10^3 \times \frac{2}{3} \times (2E)^{0.5}} = \frac{3(L - L_c)}{10^3 \times 2} \times \sqrt{\frac{m}{2E}} \tag{5}$$

When considering modern cushioned cylinders we can assume the energy capability of the cushion to be equal to or greater than 50 per cent of the maximum available thrust at a velocity of 0.5 m/s.

$$E = \frac{\tfrac{1}{2}(P \times a) \times V^2}{2g} \tag{6}$$

and, since $V = 0.5$ m/s,

$$E = \frac{\tfrac{1}{2}(P \times A) \times (0.5)^2}{2g}$$

$$= \frac{\tfrac{1}{2} \times P \times \pi D^2 \times (0.5)^2}{2g \times 4}$$

Therefore

$$E = 5 \times 10^{-4} \times P \times D^2 \; \text{N m (approx.)} \tag{7}$$

This is for a retracting cylinder (the cushion being at the full bore end); for an extending cylinder the formula would be modified to:

$$E = 5 \times 10^{-4} \times P(D^2 - d^2) \tag{8}$$

From equations (7) and (5) and rationalising constants

$$T_2 = 0.05 \times \left(\frac{L - L_c}{D^2}\right) \times \sqrt{\frac{m}{P}} \tag{9}$$

Substituting (8) and (5) and rationalising constants

$$T_2 = 0.05 \times (L - L_c) \times \sqrt{\frac{m}{P(D^2 - d^2)}} \tag{10}$$

Equations (9) and (10) represent the shortest time in which the load may be moved over the stroke prior to attaining effective cushioning.

Thus, equations (3), (9) and (10) must be evaluated to determine the longer of the stroke times.

Stage 3 (T_3)

This is the time spent within the cushion during deceleration:

$$T_3 = \frac{L_c}{V_i}$$

where V_i = maximum impact velocity (mm/s)

The value of V_i is a variable for any given cylinder application and is adjusted to suit varying loads, in order to limit impact forces. With a very lightly loaded cylinder the limiting factor on the velocity may be the flow rate through a fully open cushion, hence the manufacturer's specification should be consulted. For many cylinders a value of 300 mm/s may be adopted.

In order to assess the impact velocity V_i for a given application the following empirical formula may be considered:

$$V_i = \frac{30D}{m} \text{ (mm/s)}$$

As will be appreciated, impact velocities will vary widely with different applications so the above should be used only as a guide if no firm figure is available.

Example 4.3

A pneumatic cylinder complete with a direction control valve and associated pipe and fittings has been selected to move a mass of 3 kg with a supply pressure of 8 bar. Estimate the extend stroke time.

Solution

The manufacturer's data for the cylinder is:

$$D = 50 \text{ mm}$$

$$L = 240 \text{ mm}$$

$$L_c = 29 \text{ mm}$$

and for the valve is:

$$C_e = 2.76$$

$$T_1 = 0.05$$

Time required to reach the cushion $= T_2$.

$$T_2 = \frac{1.2(L - L_c)D^2}{C_e \times 10^6} \quad \text{from equation (3), page 117}$$

$$= \frac{1.2(240 - 29) \times 50^2}{2.76 \times 10^6}$$

$$= 0.229 \text{ s}$$

Alternatively,

$$T_{2\min} = \frac{0.05(L \times L_c) \times \sqrt{(m/P_1)}}{D}$$

$$= \frac{0.05(240 - 29) \times \sqrt{(3/8)}}{50}$$

$$= 0.129 \text{ s}$$

In this instance the time is not affected by the cushion requirements as $T_{2\min}$ is less than T_2.

From the formula $V_i = 30D/m$, the maximum cushion velocity can be estimated as:

$$V_i = \frac{30D}{m} = \frac{30 \times 50}{3} = 500 \text{ mm/s}$$

Since the velocity exceeds 300 mm/s, then 300 mm/s is adopted.

$$T_3 = \frac{L_c}{V_i} = \frac{29}{300} = 0.097 \text{ s}$$

The actual minimum stroke time $= T_1 + T_2 + T_3$:

$$0.05 + 0.229 + 0.097 = 0.376 \text{ s}$$

Example 4.4

Using the same componets as the previous example but with the load increased to 15 kg, determine the stroke time.

Solution

From the previous question

$$T_1 = 0.05 \text{ s} \quad \text{and} \quad T_2 = 0.229 \text{ s}$$

$$T_{2\,\text{min}} = \frac{0.05(L - L_c) \times \sqrt{(m/P_1)}}{D}$$

$$= \frac{0.05(240 - 29) \times \sqrt{(15/8)}}{50}$$

$$= 0.289 \text{ s}$$

Since $T_{2\,\text{min}}$ is greater than T_2 the exhaust requires to be restricted in order to provide effective cushioning. The conductance corresponding to $T_{2\,\text{min}}$ is:

$$C_e = \frac{1.2(L - L_c)D^2}{T_{2\,\text{min}} \times 10^6} \quad \text{from equation (3) page 117}$$

$$= \frac{1.2(240 - 29) \times 40^2}{0.289 \times 10^6}$$

$$= 1.402$$

Time during cushioning is:

$$T_3 = \frac{L_c}{V_i}$$

$$V_i = \frac{30D}{m} = \frac{30 \times 50}{15} = 100 \text{ mm/s}$$

Therefore,

$$T_3 = \frac{29}{100} = 0.29 \text{ s}$$

Therefore, total time is:

$$T_{\text{tot}} = T_1 + T_2 + T_3$$

$$= 0.05 + 0.289 + 0.29$$

$$= 0.629 \text{ s}$$

4.3 Cylinder mounting

When deciding on a particular form of mounting one should carefully consider the type of operation being carried out and the nature of the load. Where a load follows a straight line with little or no deviation in any plane then one of the rigid mountings shown in Fig. 4.20 may be used.

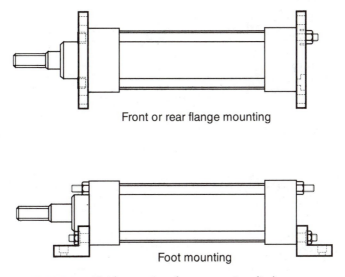

Front or rear flange mounting

Foot mounting

FIGURE 4.20 Rigid mountings for pneumatic cylinders.

Where a load is required to turn in one plane, then one of the mountings shown in Fig. 4.21 can be used.

Where the load turns in one plane but has slight movement in another plane, then swivel eyes (as shown in Fig. 4.22) can be used.

Piston rods are normally supplied with a male thread suitable for attaching either directly to a load or to a clevis arrangement.

In applications where the principal force required is to push the load, then rear mounting is usually selected. However, where the principal force required is to pull the load, front mountings will be selected. In either of these cases the stresses on the mounting bolts and cylinder body are minimised.

4.3.1 Piston rod buckling

In many pneumatic applications the buckling of the piston rod is not normally a factor to be considered. However, as stroke lengths and loads increase, the possibility of buckling occurring has to be taken into account. The piston rod will act as a strut when it is subjected to a compressive load or it exerts a thrust (Fig. 4.23). The rod must therefore be of sufficient diameter to prevent buckling.

Euler's strut theory is used to calculate a suitable piston rod diameter to withstand buckling.

Euler's formula states that:

$$K = \frac{\pi^2 \times E \times J}{L^2}$$

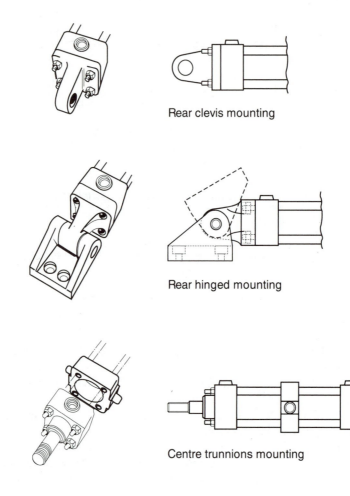

Rear clevis mounting

Rear hinged mounting

Centre trunnions mounting

FIGURE 4.21 Articulating mountings for pneumatic cylinders.

The maximum safe working thrust or load F on a piston is given by

$$F = \frac{K}{S}$$

where S is a factor of safety.

To minimise calculations, cylinder manufacturers produce charts and tables which relate piston rod diameters and operating pressures to maximum permissible buckling lengths, as shown in Fig. 4.24.

When determining piston rod diameters one should not select rod sizes at or around their theoretical limit since this will exclude any possible extra forces due to misalignment, poor installation, wear in bearings, or any unforeseen side loadings that will increase the risk of buckling.

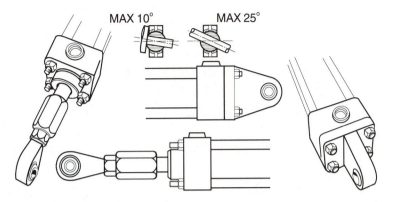

FIGURE 4.22 Swivel mountings for pneumatic cylinders.

4.4 **Rodless cylinders**

Three principal types of rodless cylinders are avaialble.

- Magnetic cylinders
- Band cylinders
- Slot-type cylinders.

4.4.1 **Magnetic rodless cylinders**

A magnetic rodless cylinder is shown in Fig. 4.25. This incorporates a series of strong magnets in both the piston and the carriage. The piston is carried within a polished stainless steel cylinder which, on the outside, becomes the slide for the carrier, which is fitted with nylon/bronze bearings.

This type of cylinder is available up to 40 mm diameter and is able to operate at speeds up to about 3 m/s. Its advantages are:

1. Internally it is not subject to contamination related wear providing the air supply is filtered.
2. It is suitable for non-lubricated air application.
3. It has almost no air leakage.
4. Carriage orientation can be set anywhere around the cylinder.
5. The extended and retracted lengths of the cylinder are the same.

The disadvantages are:

1. The carriage may become separated from the piston; however, recoupling is a simple process. This can even be an advantage in preventing overloads.
2. It may be necessary to use gaiters to protect the carrier bearings from rapid wear due to environmental contamination.
3. Higher cost than a conventional cylinder.

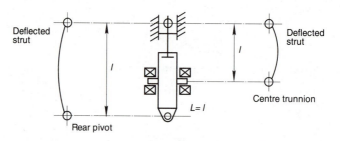

(a) Rear pivot and centre trunnion mounted. Guided pivoted load

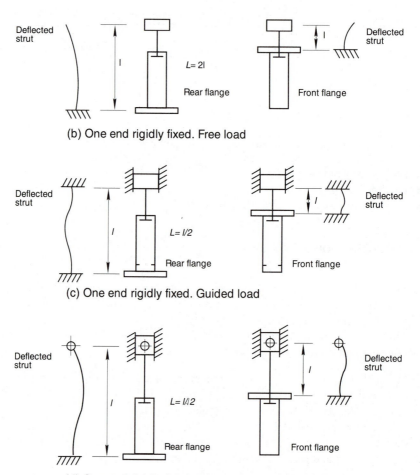

(b) One end rigidly fixed. Free load

(c) One end rigidly fixed. Guided load

(d) One end rigidly fixed. Pivoted and guided load

FIGURE 4.23 Relationship between piston rod, free buckling length (L) and method of fixing.

4.4.2 Band cylinders

Band cylinders, as shown in Fig. 4.26, use a cable or band to connect either side of the piston to the carriage. This type of cylinder is available up to 63 mm diameter and 7 m

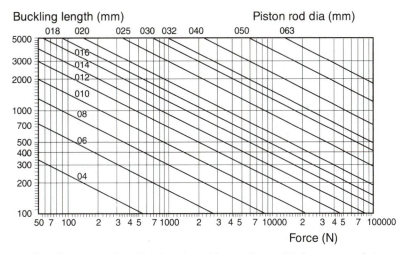

FIGURE 4.24 Chart showing the buckling length and force relationship for a range of piston sizes.

stroke and is capable of speeds up to 2 m/s with lubricated air and 1 m/s with non-lubricated air.

The principal advantage of this type of cylinder is that the piston is not subjected to misaligning forces and there is therefore minimal wear of both the bore and piston seals.

When installing band cylinders it is important to ensure the correct band tension if an automatic tensioner is not fitted. Incorrect tensioning can cause the load to move erratically and the band seals to wear excessively. For certain applications it is possible to extend the band over remote pulleys so that the load is independent of the actual cylinder, i.e. the load is carried on its own bearing supports.

4.4.3 Slot-type cylinders

These are by far the most popular type of rodless cylinders. They incorporate a minimum overall length with a certain amount of structural strength which makes them a versatile

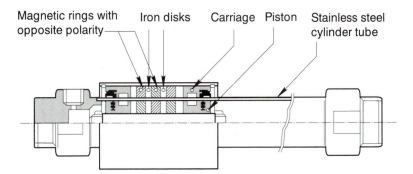

FIGURE 4.25 Typical rodless cylinder with magnetic coupling between piston and carriage.

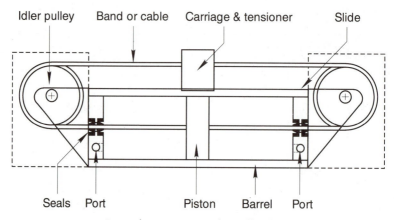

FIGURE 4.26 General arrangement of a rodless band cylinder.

solution to many problems. One of the more common applications is the operation of sliding doors in railway carriages and buses where the compact design is ideal. Slot-type cylinders are available in sizes from 16 mm bore to 80 mm bore with lengths up to 14 metres. The most critical area of consideration is the slot-sealing mechanism. As the piston and carriage traverse the cylinder, the external dust seal and the internal pressure seal must be separated to allow the tongue, connecting piston and carriage to pass. They must then be replaced in the 'at rest' sealing position. A typical slot-type cylinder is shown in Fig. 4.27.

4.4.4 Selection of rodless cylinders

When selecting any form of rodless cylinder for a particular application there are three principal factors to consider:

1. Cushioning requirements
2. Nature and magnitude of loading
3. Cylinder deflection.

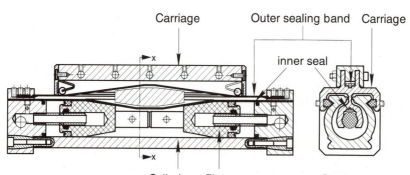

FIGURE 4.27 General arrangement of a slot-type rodless cylinder.

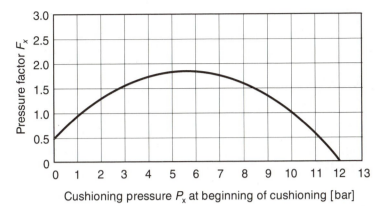

FIGURE 4.28 Range of cushioning pressure factors for a particular rodless cylinder.

Let us consider the ability of the internal cushions to dissipate the kinetic energy of the moving load and eliminate impact at the end of the stroke.

Therefore,

Cushioning energy $(F_x . E_{cush})$ = Kinetic energy of the load $(\frac{1}{2}mV^2)$

For the selection of a cylinder it is assumed that at the commencement of cushioning the exhaust pressure and the supply pressure are equal. Manufacturer's data will provide information such as cushioning energy and the pressure factor F_x (see Table 4.3 and Fig. 4.28).

TABLE 4.3 Cushioning energy potential for a typical range of rodless cylinders.

Cylinder dia. (mm)	Cushioning energy E_{cush} (N m)	Sectional area A (cm^2)
25	2.1	4.9
32	4.0	8.0
40	8.7	12.5
50	14.4	19.6
63	29.8	31.1

Hence,

$$E = \frac{mV^2}{2} \geqslant F_x . E_{cush}$$

where E = kinetic energy (N)
   m = mass (kg)
   V = velocity (m/s)
   F_x = pressure factor (see Fig. 4.28)
   E_{cush} = cushioning energy

When considering braking one must take into account the exhausting pressure on the piston and the modes of operation of the cylinder, as shown in Fig. 4.29.

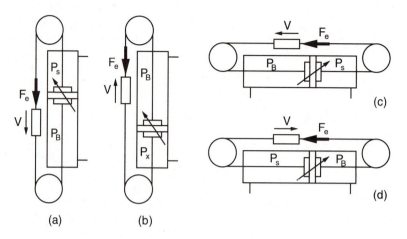

FIGURE 4.29 Orientation of a rodless cylinder.

In order to determine the actual cushioning pressure it is necessary to examine the mode of operation.

For modes (a) and (c) the following empirical equation applies:

$$P_{cush} = P_s + \frac{10 \times F_e}{a}$$

$$= \text{Maximum cylinder cushioning pressure}$$

alternatively, for modes (b) and (d)

$$P_{cush} = P_s - 10 \times \frac{F_e}{a}$$

$$= \text{Maximum cylinder cushioning pressure}$$

where P_{cush} = cushioning pressure at the start of cushioning (bar)
   P_s = supply pressure (bar)
   F_e = external load (N)
   a = cross-sectional area of cylinder (mm²)
   10 = constant

Example 4.5

In a vertically mounted cylinder application a mass of 5 kg is driven downwards at 1.5 m/s by a supply pressure of 8 bar.

The maximum working pressure of the cylinder is 10 bar. Determine a cylinder size that will provide adequate internal cushioning.

Solution

$$E = \frac{mV^2}{2} = F_x . E_{cush}$$

where $m = 5$ kg
$V = 1.5$ m/s
$F_x = 1.5$ (from Fig. 4.28)

Therefore,

$$E = \frac{5 \times 1.5^2}{2} = 1.5 \times E_{cush}$$

$$E_{cush} = \frac{5 \times 1.5^2}{2 \times 1.5}$$

$$= 3.75 \text{ N m}$$

Cylinder selection from Table 4.3 shows that, 32 mm diameter has 4.0 N m E_{cush} potential.

Determination of cushioning pressure (P_{cush}) is

$$P_{cush} = P_s + 10 \left(\frac{F_e}{a} \right)$$

where F_e = Mass $\times$ Acceleration due to gravity
$= 5$ kg $\times$ 9.81 m/s^2
$= 49.05$ N

Therefore,

$$P_{cush} = 8 + \frac{10 \times 49.05}{800}$$

$$= 8.6 \text{ bar}$$

Hence, since the cushioning pressure is less than the maximum working pressure, a 32 mm bore cylinder can be braked effectively.

Example 4.6

In a vertically mounted cylinder application a mass of 24 kg is driven upwards at 1.8 m/s by a supply pressure of 6 bar. If the maximum working pressure is 10 bar determine a cylinder size that will provide adequate internal cushioning.

Solution

$$E = \frac{mV^2}{2} = F_x . E_{cush}$$

where $m = 24$ kg
   $V = 1.8$ m/s
   $F_x = 1.8$ (from Fig. 4.28)

Therefore,

$$E_{cush} = \frac{mV^2}{2F_x}$$
$$= \frac{24 \times 1.8^2}{2 \times 1.8}$$
$$= 21.6 \text{ N m}$$

Cylinder selection from Table 4.3 shows that a 63 mm diameter cylinder has 29.8 N m E_{cush} potential.
   Determination of cushioning pressure (P_{cush}) is:

$$P_{cush} = P_s - 10 \left(\frac{F_e}{a}\right)$$

where F_e = Mass × Acceleration due to gravity
   $= 24 \times 9.81$ N
   $= 235.44$ N

Therefore,

$$P_{cush} = 6 - \frac{10 \times 235.44}{3110}$$
$$= 6 - 0.76$$
$$= 5.24 \text{ bar}$$

Calculation confirms that a 63 mm bore cylinder is suitable.

4.4.5 Cylinder deflection

The majority of cylinders are designed to tolerate a deflection of up to 2 mm without adverse effect. It is therefore important to determine the position of the cylinder supports so that the maximum deflection is not exceeded. Figure 4.30 shows a selection of support arrangements that can be applied to appropriate installations.

   Manufacturers will provide graphs to illustrate loads and their relative support lengths in order to make determination of support lengths simple. Figure 4.31 gives a typical example. However, it may be necessary to calculate the actual distances between supports and the number of supports required.

Calculating the minimum number of pitches

$$t_{min} = \frac{A_1 + stroke}{L_1}$$

where t_{min} = minimum number of pitches
   A_1 = minimum distance from the centre of the carriage to the end of the
      cylinder (mm)

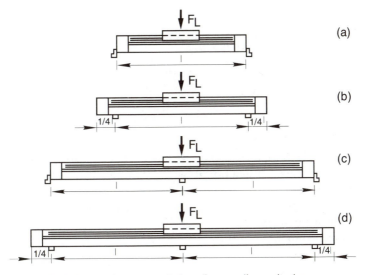

FIGURE 4.30 Support variations for a rodless cylinder.

stroke = stroke of cylinder (mm)

L_1 = maximum distance between supports (from deflective charts, Fig. 4.31) (mm)

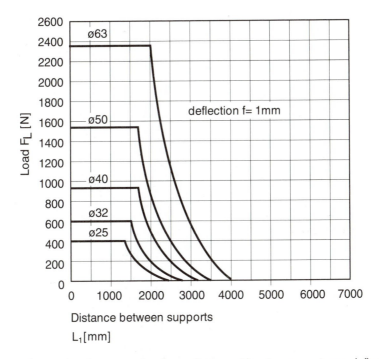

FIGURE 4.31 Load capacity of a range of rodless cylinders with a 1 mm maximum deflection.

Calculation of pitch

Determination of the pitch will be dependent upon which method of support is selected from Fig. 4.30. In all cases the following formula can be applied:

$$L = \frac{A_1 + \text{Stroke}}{t_{sel}}$$

where L = actual distance between supports (mm)
   t_{sel} = selected pitch $(t_{sel} > t_{min})$.

Depending upon which support method is chosen from Fig. 4.30, t_{sel} is rounded up to the nearest whole number if support (a) or (c) is selected or rounded up to the nearest half is support (b) or (d) is selected.

Hence the number of supports N required is derived as follows:
Mounting method (a) or (c): $N = t_{sel} + 1$.
Mounting method (b) or (d): $N = t_{sel} + 0.5$.

Example 4.7

A 50 mm diameter cylinder, having a stroke of 5000 mm, is to carry an external load of 1400 N. If the maximum deflection is not to exceed 1 mm, determine the number and pitch of the supports.

Solution

A_1 is determined from manufacturer's data. In this case we will use 300 mm.

$$t_{min} = \frac{A_1 + \text{Stroke}}{L_1}$$

$$= \frac{300 + 5000}{1700}$$

$$= 3.18$$

Actual length between pitches, support method (c):

$$L = \frac{A_1 + \text{Stroke}}{t_{sel}}$$

t_{sel} for support method (c) = 4; and t_{sel} for support method (d) = 4.5. Thus,

$$L = \frac{300 + 5000}{4}$$

$$= 1325 \text{ mm}$$

Actual length between pitches, support method (d):

$$L = \frac{A_1 + \text{Stroke}}{t_{sel}}$$

$$= \frac{300 + 5000}{4.5}$$

$$= 1178 \text{ mm}$$

Since both alternatives are below the maximum value of that stated in Fig. 4.31, i.e. 1700 mm, both are acceptable methods of support.

4.5 Cylinder seals

The 'O' ring and quadring are used where pistons travel at relatively low speeds and operate at low pressure. With the 'O' ring the risk of seal extrusion may necessitate the use of a backing ring to prevent extrusion.

Lip-type seals offer superior characteristics both at low and high pressure. When installed the pre-tensioned lip seal holds itself in contact with one of the surfaces to be sealed while the base and shoulder of the seal contacts the other surfaces to be sealed. As pressure is applied to the seal, it increases the seal spread and improves its sealing characteristics. The ability of the seal to spread also allows it to compensate for any wear that takes place either to the seal or to the moving surface.

A variety of materials and seal profiles are available with an operating temperature range of -25 to $+80\,°\text{C}$.

4.6 Torque units (semi-rotary actuators)

Where movement through a specific angle is required it is possible to use standard double-acting cylinders and levers. However, this is not normally a satisfactory method since it is bulky, restricted to relatively small angles and the output torque varies throughout the piston stroke.

Given these disadvantages the semi-rotary actuator is an ideal alternative, and can be a piston type or a vane type. The piston type consists of a piston rod in the form of a rack with a piston at either end, and a pinion which carries the output drive shaft is in mesh with the rack, as shown in Fig. 4.32.

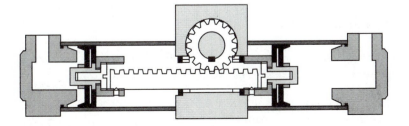

FIGURE 4.32 General arrangement of a piston-type semi-rotary actuator.

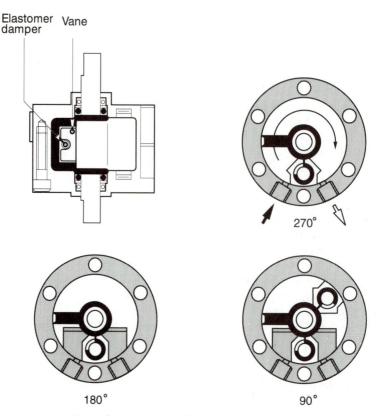

Elastomer damper Vane

270°

180°   90°

FIGURE 4.33 General arrangement of a vane-type semi-rotary actuator.

With this type of actuator equal forces and speeds can be easily applied in either direction of rotation. These units are usually available for angular movement up to 360° in 90° increments, with output torques up to 400 N m.

The vane type semi-rotary actuators shown in Fig. 4.33 have lower torque outputs and are limited to about 280° of rotation for a single vane and 100° for a double vane.

Semi-rotary actuators offer effective systems for operating process valves, for manipulating workpieces and for robotics applications.

4.7 **Pneumatic motors**

The pneumatic motor forms a very viable alternative to both electric and hydraulic motors in the power range up to 25 kW. Its principal advantages are:

1. High power to weight/size ratio
2. Can be stalled without damage
3. Simple speed control
4. Suitable for use in hazardous areas

Type	Displacement motors			
Characteristic	Radial piston motor	Linked piston motor	Vane motor	Gear motor
		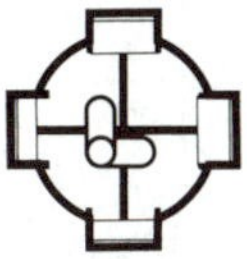		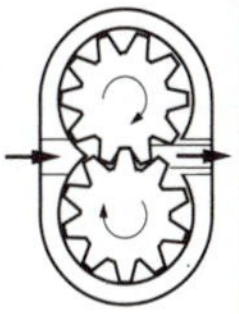
Max. operating pressure (bar)	10	8	8	10
Rated power (kW)	1.5 to 30	1 to 6	0.1 to 18	0.5 to 5
Max. speed (rpm)	6000	5000	30000	15000
Specific air consumption (1/kJ)	15 to 23	20 to 25	20 to 50	30 to 50
Lubrication	Sump and/or with compressed air	Sump and/or with compressed air	Compressed air	Compressed air

FIGURE 4.34 Motor Comparisons.

5. Easy to reverse
6. Extremely robust
7. Minimal maintenance
8. Simple installation
9. Torque remains fairly constant across a wide range of speeds
10. Capable of very high speeds.

Pneumatic motors can be divided into two categories: (1) displacement motors in which the actuator chamber changes its size during operation; (2) dynamic motors in which the actuator chamber remains unchanged during operation.

4.7.1 Displacement motors

The principal types of displacement motors are shown in Fig. 4.34.

4.7.2 Vane motors

These are probably the most popular and widely used of the displacement motors. They offer the most compact and versatile construction and are generally available in powers up to 5 kW with speeds ranging from 200 to over 30 000 rev/min.

Figure 4.35 shows the internal construction of a typical vane motor. The core of the motor is a simple design, comprising only a few components. A slotted rotor rotates eccentrically within the chamber formed by the cylinder and end plates. Since the rotor is smaller in diameter and eccentric to the outer chamber, a crescent-shaped chamber is formed. The rotor is provided with vanes which are allowed to move freely in and out

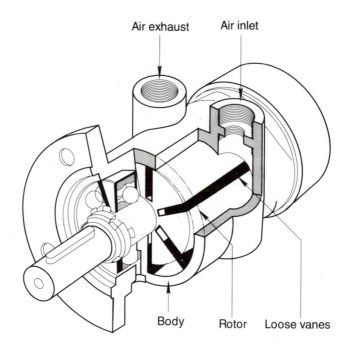

FIGURE 4.35 General arrangement of a vane motor.

of the slots in the rotor. These vanes divide the chamber into separate compartments, each of a different size. Compressed air is admitted into the compartments at a point where its volume is enlarging. The air pressure acting on the inlet side of the vane creates a force, and hence a torque or turning moment results. The actual quantity of air supplied determines the speed of the motor. In vane motors the actual expansion of the air is responsible for approximately 20% of the total energy of the motor.

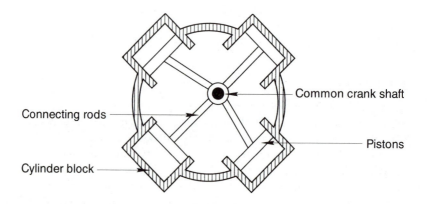

FIGURE 4.36 General arrangement of a radial piston motor.

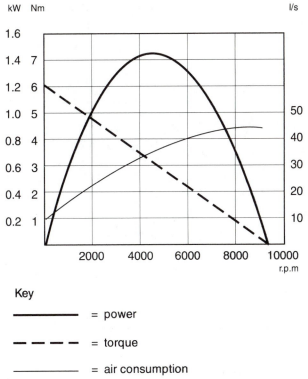

FIGURE 4.37 Typical pneumatic motor characteristics.

Under start-up conditions and slow-speed operation the pressurised air flows behind the vanes and pushes them out against the cylinder wall, thereby creating a seal between each chamber. As the rotational speed increases the centrifugal effect provides the necessary sealing force and at very high speeds these forces can be excessive, thereby increasing the wear rate of the motor. The wear rate of the vane tip and the cylinder generally limits the motor speed.

Where very high speed motors are required the motor design is modified to provide longer, thinner vanes, thereby reducing vane forces to an acceptable level.

4.7.3 Piston motors

Piston motors are designed to provide greater output powers than vane motors, up to 25 kW at speeds of between 2000 and 5000 rev/min. Piston motors similar to those shown in Fig. 4.36 offer some advantages over vane motors. Owing to a higher volumetric efficiency, speed control is much easier. They are also easily reversed by simply reversing the direction of the air flow.

The principal disadvantages of radial piston motors are that they are much bulkier and more complex in construction than vane motors.

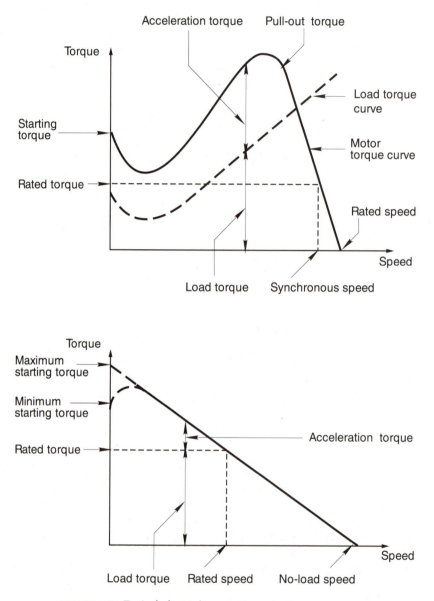

FIGURE 4.38 Typical electric/pneumatic motor torque curves.

Axial piston motors such as the swash plate motor offer a half-way stage between the radial piston motor and the vane motor. This type of motor is smaller than the radial piston motor but can still produce relatively high torques. One of the principal advantages of this type of motor is that its speed can be varied without altering the air supply. This is brought about by varying the angle of the swash plate; as the angle of the swash plate is decreased there is an increase in speed but a corresponding decrease in torque and vice versa.

4.7.4 Dynamic motors

Gear motors

The gear motor generally has three different gear forms:

straight teeth, helical teeth and double helical teeth. These motors are purely displacement motors and do not use the expansive properties of compressed air. Gear motors are used mainly in heavy industrial applications for hazardous areas where electric motors would not be allowed and are capable of operating to powers in excess of 300 kW. The gear motor has a lower volumetric efficiency than either the piston or vane motor.

4.7.5 Motor characteristics

Comparing the characteristics of a pneumatic motor to those of an electric motor (Fig. 4.37) it can be seen that the torque/speed curve for a pneumatic motor is almost a straight line compared with the steep curve of the electric motor. Also, the operational speed range of the pneumatic motor is much greater than that of the electric motor.

Figure 4.38 shows a typical graph relating all the characteristics of a pneumatic motor.

The graph shows that the maximum power output of the motor occurs at approximately half the maximum no-load speed. It is the speed relative to the maximum power at which the motor is rated. This type of graph is usually produced for operating pressures between 6 and 7 bar gauge.

For any pneumatic motor the torque is a function of the input pressure, while the speed is a function of the quantity of air flowing. One can, therefore, make adjustments to both speed and torque very easily. If one adjusts the pressure regulator on the supply then the potential output torque is modified.

CHAPTER 5

Cylinder control

5.1 Direction control

Although the majority of pneumatic circuits have more than one cylinder it is important that the methods by which one cylinder is controlled be fully understood since the control techniques can then be applied to multi-cylinder circuits.

A cylinder rod is extended or retracted by directing a pressurised flow of air to one side of the piston and exhausting the other. This chapter is concerned with the various ways of controlling the motion of the cylinder piston rod.

5.1.1 Single-acting cylinders

A single-acting spring-retracted cylinder controlled by a 3/2 spring offset direction control valve is shown in Fig. 5.1(a) in the 'at rest' condition. When the valve is operated, as shown in Fig. 5.1(b), air flows to the full bore side of the piston and the piston rod extends.

If the ports on the valve connected to the air supply and exhaust were reversed, the piston rod would be extended when the valve was not being operated, as shown in Fig. 5.2. Operating the valve would exhaust the air in the full bore side allowing the spring force to retract the piston rod.

The valves shown in Figs. 5.1 and 5.2 are both 3/2 valves, but one is normally closed and the other normally open. In some direction control valve designs the inlet and outlet ports are interchangeable but there are exceptions and the suitability of the component for non-standard operation must be checked.

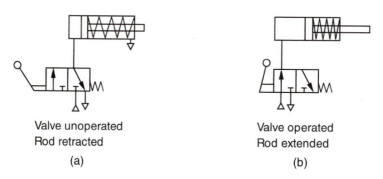

Valve unoperated
Rod retracted
(a)

Valve operated
Rod extended
(b)

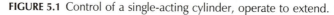

FIGURE 5.1 Control of a single-acting cylinder, operate to extend.

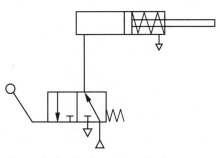

FIGURE 5.2 Control of a single-acting cylinder, operate to retract.

5.12 Double-acting cylinders

To control the extend and retract strokes of a double-acting cylinder, the cylinder ports have to be pressurised and exhausted. The supply of air to and from the cylinder uses four 2/2 valves, two 3/2 valves, one 5/2 or one 5/3 valve. The use of 3/2 valves to control a cylinder is shown in Fig. 5.3

With neither valve operated, both sides of the piston are exhausted so the piston rod is free to move in either direction. If valve V_A is operated alone the piston rod extends and, conversely, if valve V_B only is operated the piston rod will retract. If, however, both valves are operated at the same time both sides of the piston will be pressurised, the larger force being exerted on the full bore side as it has the larger area; so, depending upon friction and cylinder load, the piston will try to extend.

Two 3/2 valves are only used to control cylinders in special circumstances, e.g. with a very long stroke cylinder the individual valves may be located near to each port, minimising pipe runs between the valve and the cylinder thereby reducing air consumption. It is, however, far more common to use 5/2 valves and occasionally 5/3 valves. A double-acting cylinder controlled by a 5/2 spring offset push-button-operated valve is shown in Fig. 5.4.

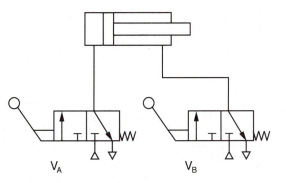

FIGURE 5.3 Control of a double-acting cylinder, with individual 3-port valves.

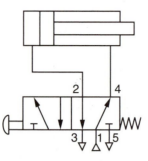

FIGURE 5.4 Control of a double-acting cylinder, with one 5-port, 2-position valve.

With the valve in the unoperated or at-rest condition, supply port 1 is connected to service port 4, and service port 2 is connected to exhaust port 3, and the piston is retracted. When the push-button is operated, port 1 connects to port 2 and port 4 connects to port 3 and the cylinder extends.

Using a 5/2 valve the cylinder only has two at-rest conditions: fully extended or fully retracted. A 5/3 valve can be utilised to give an intermediate condition, as shown in Fig. 5.5(a) and (b).

The circuit in Fig. 5.5(a) shows a three-position valve with both service ports connected to exhaust in the mid-position which allows the piston to 'float' or 'free wheel'. This configuration of a 5/3 valve is sometimes referred to as an 'open centre'.

The circuit shown in Fig. 5.5(b) shows a blocked or closed centre 5/3 valve. This valve may be used under emergency shutdown conditions as it may hold the piston of a double-acting cylinder in an intermediate position with pressurised air trapped on both sides of the piston. Accurate positioning is not possible because when the valve shown in Fig. 5.5(b) is centred the piston will continue to move, compressing air on one side of the piston, and allowing it to expand on the other. This movement will continue until the forces on both sides of the piston are in balance. Exhaust restrictors, which set up a back pressure, will reduce the movement of the piston prior to the force balance being achieved whenever the valve is centralised.

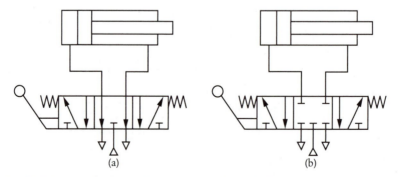

(a)   (b)

FIGURE 5.5 Control of a double-acting cylinder, with 5-port, 3-position valves.

Example 5.1

A wire basket containing red hot steel components is to be lowered slowly into an oil quench bath, as shown in Fig. 5.6. The basket has to be stopped and held in any position, the baskets stopping accuracy is unimportant. Devise a pneumatic circuit and estimate the bore of a suitable pneumatic cylinder to ISO recommended sizes. The total basket weight with components is 60 kg and the maximum air supply pressure available is 5 bar gauge.

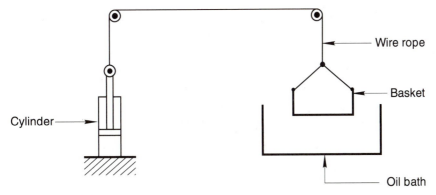

Cylinder

Wire rope

Basket

Oil bath

FIGURE.5.6 General arrangement of the quenching system.

Solution

Use a 5/3 closed centre valve with exhaust restrictors, as shown in Fig. 5.7. The

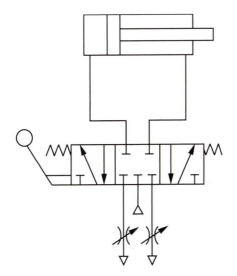

FIGURE 5.7 Circuit diagram to operate the quenching system.

exhaust restrictors are adjusted so that a fully loaded basket will just raise or lower when the valve is operated. In sizing the cylinder it must be noted that there will be a considerable back pressure; and in a case such as this when the cylinder is extending slowly with the load assisting the motion, the back pressure will be greater than the supply pressure. When the wire basket is being raised, the load and the back pressure oppose the motion of the cylinder, and the cylinder size must be calculated under this condition. The retract thrust of the cylinder must overcome the load and the back pressure (see Fig. 5.8).

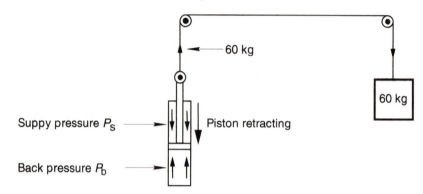

FIGURE 5.8 Representation of forces when retracting the cylinder.

If the full bore area of the piston is A, the rod area is a, the air supply pressure is P_s and the back pressure is P_b, then equating forces on the piston:

$$P_b \times A + 60 \text{ kg} = P_s(A - a)$$

When the exhaust restrictor is adjusted it will vary the back pressure P_b. If we assume that P_b is 60 per cent of the supply pressure, then

$$P_b = 0.6P \quad (\text{where } P = 5 \text{ bar})$$

$$= 0.6 \times 5 \times 10^5 \text{ N/m}^2$$

$$\text{Note that } 60 \text{ kg} = 60 \times 9.81 \text{ N}$$

Let the units for A and a be m². Then equating forces across the piston, assuming no acceleration,

$$0.6 \times 5 \times 10^5 \times A + 60 \times 9.81 = 5 \times 10^5(A - a)$$

The rod area of a standard pneumatic cylinder is small compared with full bore area and may be neglected in this case. Then

$$0.6 \times 5 \times 10^5 \times A + 60 \times 9.81 = 5 \times 10^5 \times A$$

$$60 \times 9.81 = 0.4 \times 5 \times 10^5 \times A$$

$$A = \frac{60 \times 9.81}{2 \times 10^5}$$

If the cylinder bore is D, then $A = \pi D^2/4$, or

$$\frac{\pi D^2}{4} = A = \frac{60 \times 9.81}{2 \times 10^5} \text{ m}^2$$

Thus,

$$D^2 = \frac{4 \times 60 \times 9.81}{\pi \times 2 \times 10^5} \text{ m}^2$$

$$D = \sqrt{\frac{4 \times 60 \times 9.81}{\pi \times 2 \times 10^5}} \text{ m}$$

$$= 62 \times 10^{-3} \text{ m}$$

$$= 62 \text{ mm}$$

The nearest standard cylinder is 63 mm bore, which is very close to that calculated and as the rod diameter has been neglected it would be advisable to use the next size higher, i.e. an 80 mm bore cylinder. A 63 mm bore cylinder could be used if the exhaust restrictor was adjusted to reduce the back pressure to a value below 60 per cent of the supply pressure.

5.2 Speed control

The speed of a pneumatic actuator depends upon the volume of air entering and leaving the unit. Air is a compressible fluid and so the volume of a given mass of air depends

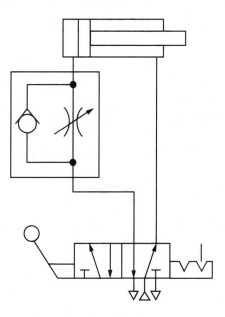

FIGURE 5.9 Meter-in speed control.

upon its pressure and temperature. The change of temperature of the air across an actuator is normally very small and may be neglected; however, the pressure can vary greatly as the load and flow rates change. A degree of speed control can be achieved by using restrictors or flow control valves to meter the quantity entering (meter-in flow control) or leaving (meter-out flow control) an actuator. Should precision speed control be required, some other form of regulator must be used, typically some form of hydraulic control, as described later in the chapter.

5.2.1 Meter-in

In this case the flow of air entering the cylinder is controlled by a variable restrictor, as shown in Fig. 5.9.

A non-return valve in parallel across the restrictor allows free flow of air in the reverse direction.

Meter-in speed control is not recommended for pneumatic systems with variable loads when good speed regulation is required. Should the load overrun, the system will lose control.

5.2.2 Meter-out

In this method of speed control the exhaust air is metered, which sets up a back pressure that will compensate to some degree for variations in the cylinder load. Meter-out speed control can be achieved either by using a flow control valve and non-return valve in the line between the cylinder and the direction control valve or by using a restrictor fitted into the exhaust port of the direction control valve as shown in Fig. 5.10.

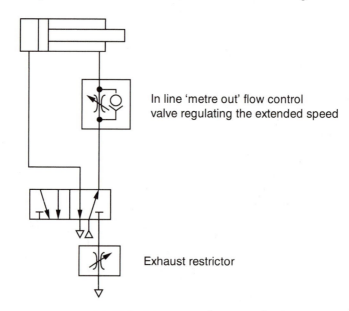

In line 'metre out' flow control valve regulating the extended speed

Exhaust restrictor

FIGURE 5.10 Meter-out flow control.

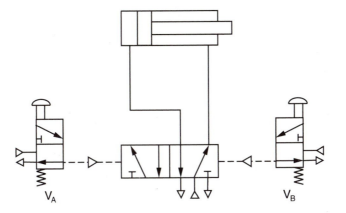

FIGURE 5.11 Operation of a cylinder using remote pilot control valves.

An 'in-line' flow control valve gives a more exact control of the air flow than an exhaust port restrictor and should be used where fine control is required.

5.3 Pilot operation

Valves used to drive pneumatic actuators have to be correctly sized to handle the required air-flow rate. When large valves are used the pipework connecting the valves and actuators must be of a correspondingly large bore. The valves should be positioned as near as possible to the actuator so as to reduce the length of pipework between the valve and actuator to a minimum; this will result in a reduced air consumption per cycle. The valve may have to be operated remotely, and one method of doing this is to use a pilot air signal to operate the main valve spool.

5.3.1 Pressure-applied operation

A circuit using 3/2 push-button pilot valves to control the main 5/2 valve is shown in Fig. 5.11. This circuit uses the application of a pilot pressure to switch the main valve.

The circuit shown is bistable, the piston rod extends when valve V_A is operated, and remains extended even when valve V_A is released. The piston retracts when valve V_B is operated.

There will be a time delay between the pilot valve V_A operating and the piston rod extending; this is partly due to the time taken in pressurising the pilot line and depends upon the volume of the line and the force to move the valve. This time delay can be reduced by using small bore pilot lines and high pilot pressures. If very fast operation is needed the use of electrical signalling must be considered. Should valves V_A and V_B be operated simultaneously, both pilot lines will be pressurised and the main valve will remain in the same condition as it was before the pilot valves were operated. The operation of both pilot valves gives rise to a 'trapped' signal, so called because it traps the main valve in position preventing any movement.

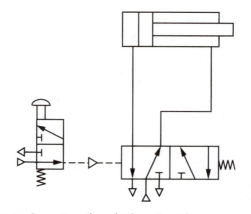

FIGURE 5.12 Operation of a cylinder using pilot pressure release.

5.3.2 Pressure release pilot operation

An example of pilot pressure release operation is shown in Fig. 5.12.

The pilot line is pressurised when the pilot valve is not operated, and the cylinder is power retracted. When the pilot valve is operated, the pilot pressure is released, the main valve moves across under the action of the reset spring, and the piston rod is extended. This is a monostable circuit, the cylinder always being in the retract condition when the pilot valve is in the unoperated condition.

Pressure release pilot operation should be used with great caution, as a failure in pilot pressure due to a fractured pipe, a loose coupling, etc., will cause the cylinder to operate. Pressure release operation can be incorporated into circuits as a safety feature, as is shown in Fig. 5.13.

Under normal conditions the safety valve is piloted across and the main air supply connects directly to the system. Should there be an air supply failure, the safety valve resets and the emergency supply receiver is connected to the system to ensure that an operation can be completed.

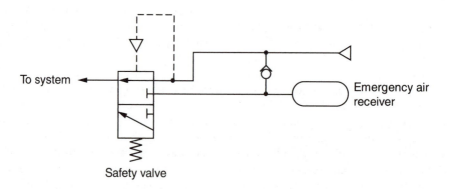

FIGURE 5.13 Arrangement for emergency air supply.

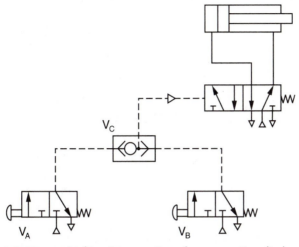

FIGURE 5.14 Multi-point operation of a pneumatic cylinder.

5.3.3 Multi-point pilot operation

There are many instances where an actuator has to be controlled from more than one position. One method of multi-point operation is shown in Fig. 5.14, in which a pilot signal from either valve V_A or V_B will pass through the shuttle valve V_C and operate the main direction control valve.

5.3.4 Automatic operation

A sensor can be incorporated into a circuit to detect the stoppage or completion of an actuator movement. The activation of the sensor is then used to initiate the next step of the sequence. The sensors may be mechanical, pneumatic, electrical, etc. Pneumatic trip valves operated either directly or indirectly by the piston rod are often used to generate a pilot signal. The circuit in Fig. 5.15 shows an automatic return system where the cylinder will not retract until the trip valve is operated at the end of the extend stroke.

If the push-button valve V_A is held in the operated condition the cylinder will stay extended although valve V_B will have been operated as there will be a trapped signal on the ' + ' pilot of the main valve. Should this be an undesirable feature, a second trip valve can be fitted to detect when the cylinder is fully retracted, as shown in Fig. 5.16.

When the stop/run valve V_A is in the stop condition as shown, the cylinder will be retracted; consequently, valve V_C will be operated and must be shown thus. When valve V_A is in the run condition, a pilot signal passes from valve V_C through valve V_A to the ' + ' pilot side of the main valve, which will switch over, causing the piston rod to extend. At this point valve V_C is released and the extend (+) pilot signal is exhausted, the piston rod fully extends and trip valve V_B is operated, sending a pilot signal to the retract (–) pilot port of the main valve, retracting the piston rod. If the stop/run valve V_A is held in the run condition, the cylinder will continue cycling until valve V_A is switched to the

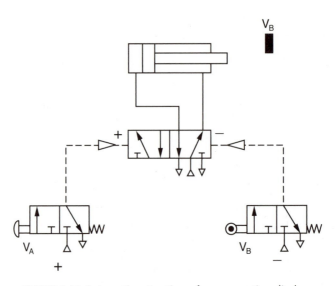

FIGURE 5.15 Automatic retraction of a pneumatic cylinder.

stop position, the cylinder will then complete its cycle and stop with the piston rod fully retracted.

5.3.5 Pressure sensing

In some cylinder applications the piston does not always extend to the same point, e.g. in clamping operations. In such cases a trip valve, which detects position, cannot be used and an alternative method of sensing must be found. The variations in the pressure of the air across the piston of a pneumatic cylinder during its extend stroke are shown against a time base in Fig. 5.17.

The diagram shows that there is a delay between the control valve being operated and the start of piston movement. This time delay depends upon the rate of change of pressure across the piston. It is influenced by the length and bore of the pipework between the valve and the cylinder, and by any flow control valves or restrictors in the circuit. Variations in the applied load will also influence the shape of the curve.

With the piston rod extending, the pressure on the full bore side of the piston only reaches full system pressure when the piston has stopped moving. This pressure can be sensed using a pneumatic sequence valve or an electrical pressure switch. Similarly, the pressure on the annulus side of the piston decays to atmospheric pressure after completion of the extend stroke, and this low pressure can be sensed using a diaphragm valve or a pressure switch. The circuit shown in Fig. 5.18 illustrates both high- and low-pressure sensing to give automatic reciprocation of a piston.

With the stop/run valve in the off position, the piston will be retracted and the annulus side at full system pressure. The sequence valve will be operated, supplying pressurised air to the stop/run valve; the diaphragm valve will also be in an operated condition causing the ' − ' retract port on the 5/2 valve to be exhausted. When the

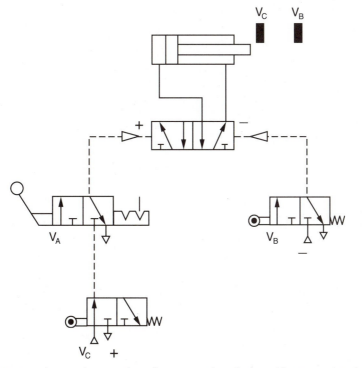

FIGURE 5.16 Automatic retraction of a pneumatic cylinder with start-up interlock.

stop/run valve is moved into the run position, a signal is applied to the ' + ' port of the 5/2 valve causing it to move and the piston to extend; the sequence valve is de-energised and the pressure in the annulus side of the cylinder and pipework falls to a very low value as the piston stops. The diaphragm valve now resets and provides a ' − ' signal to the 5/2 valve, retracting the piston.

Two sequence valves or two diaphragm valves could be used to give pressure-sensing operations on both the extend and retract strokes of the cylinder.

5.3.6 Trapped signals and signal breakers

As already mentioned a trapped signal occurs when both pilot signals exist on opposite ends of a valve simultaneously. In some applications it is essential to prevent this occurring and so the trapped signal must be eliminated. One method is to use a signal breaker valve, as shown in Fig. 5.19.

When valve V_A is operated a signal passes through the signal breaker to the ' + ' side of the 5/2 valve and the piston extends. If the push-button valve V_A is held down, the pilot signal will operate the signal breaker and cause the pilot signal from V_A to be isolated while simultaneouly venting the trapped ' + ' signal. The purpose of the restrictor is to ensure that the signal reaches the 5/2 valve before the signal breaker is switched. The signal breaker will remain energised until valve V_A is released.

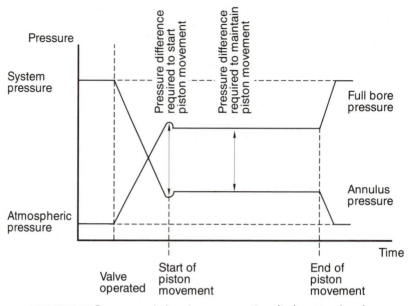

FIGURE 5.17 Pressure variations in a pneumatic cylinder extend cycle.

Commercial signal breakers are available as complete units and are known as pulse-generating valves.

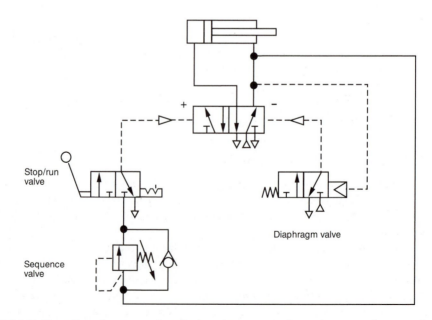

FIGURE 5.18 Operation of a pneumatic cylinder using pressure rise and pressure decay sensing.

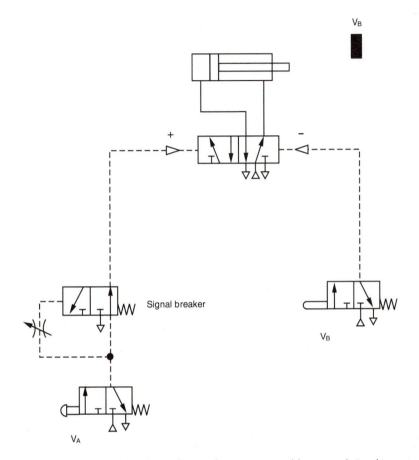

FIGURE 5.19 Signal breaker valve to eliminate a possible trapped signal.

In multi-cylinder circuits, 3/2 double-pilot-operated valves can be used to switch trapped signals off and on. A switch valve arrangement is shown in Fig. 5.20. The set and reset signals to the switch valve are taken from points in the circuit so that the switch valve is 'set' before the ' + ' signal is received and 're-set' before the ' − ' signal occurs.

5.3.7 Time delays

Time delays are used to keep a piston in one position for a period of time after the signal has been sent to cause the piston to move. In the circuit shown in Fig. 5.21 the piston will remain in the extended position for a time after the retract signal has been sent. The value of the time delay can be adjusted by varying the flow restrictor.

The actual time delay depends upon the air pressure in the reservoir rising sufficiently to overcome the friction force within the 5/2 valve. The actual pressure needed to switch the 5/2 valve will depend on valve lubrication, seal condition, valve temperature and air pressure. The accuracy of this type of pneumatic time delay could only be guaranteed to

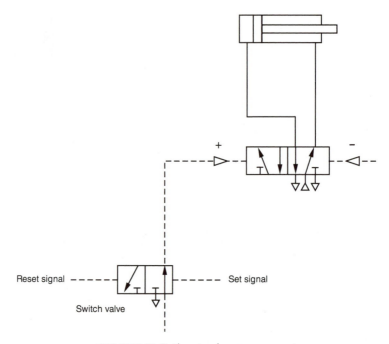

FIGURE 5.20 Set/reset valve arrangement.

within about 10 per cent; when more accurate time delays are needed commercially available pneumatic timers or electronic timers are recommended.

5.3.8 Thrust control

The thrust developed by a pneumatic piston depends upon its area and the applied pressure plus the effect of any back pressure. To control the thrust a pressure regulator is used to adjust the supply pressure. A pressure regulator can be used to control the whole circuit or an individual cylinder, as shown in Fig. 5.22.

The pressure regulator used in Fig. 5.22 sets the supply pressure for both the extend and retract strokes to a value below the main supply pressure. This will result in a lower thrust and a reduced air consumption per stroke in terms of free air used. When large-bore, long-stroke cylinders are used the air pressure needed to retract the piston may be considerably lower than the pressure required for the extend stroke. In such cases there can be a considerable saving in free air consumption by regulating both the extend and retract air pressures to the desired value. This can be done as shown in Fig. 5.23.

Note: In this application ports 1 and 5 are being used for the air supply, therefore one must ensure that the valve is suitable for back porting.

5.3.9 Manual overrides

It is often desirable to have manual controls included in automatic sequences for either setting up the system or as a safety override.

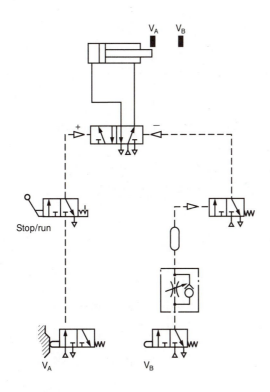

FIGURE 5.21 Time delay valve.

Consider the continuous sequence shown in Fig. 5.15. This circuit can be modified to give manual control of either extension or retraction of the piston, as shown in Fig. 5.24.

The selector valve directs a pilot air supply to either of the piston trip valves (a_0 and a_1) or to the manual extend and retract valves. Shuttle valves (s_1 and s_2) act as logic OR gates and allow the pilot signal from either the trip valves or the manual valves to operate the main power valve V_A.

5.4 Sequential control of actuators

A sequence may be:

(a) event based
(b) time based
(c) a combination of event and time based.

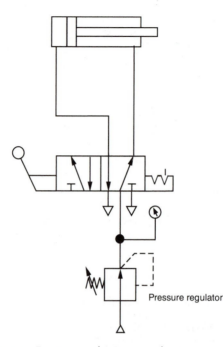

FIGURE 5.22 Force control using a single pressure regulator.

In an event-based sequence the completion of one operation causes the next operation to start. Time-based sequences have events occurring at pre-set time intervals and are usually controlled by either mechanical or electronic programmers. In this chapter only event-based sequences or event-based with time delays will be considered.

5.4.1 Notation

Each cylinder will be given a reference letter A, B, C, etc. The cylinder state, i.e. extended or retracted, is denoted by + or −, thus A+ means that cylinder A is extended and A− means that cylinder A is retracted. Consider the sequence given as A+ B+ A− B−. This means:

Start of cycle
   Then cylinder A extends (A+)
   Next cylinder B extends (B+)
   Then cylinder A retracts (A−)
   Finally, cylinder B retracts (B−)
End of cycle.

When a cycle is continuous, i.e. repeating until it is switched off, it may be shown as

$$A+ \ B+ \ A- \ B- \ repeat$$

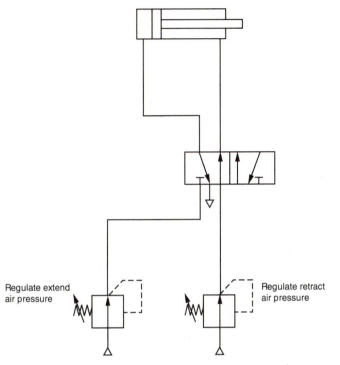

FIGURE 5.23 Dual force control using two pressure regulators.

The way in which the sequence is written gives no indication of the way in which the sequence is controlled. Trip valves which are used to indicate when the cylinder has reached the extremes of the stroke will be referred to as a_0, a_1, b_0, b_1, and so on. Trip valve a_0 being located at the retract position of cylinder A and trip valve a_1 at its extended position.

5.4.2 Two-cylinder operation

Consider the sequence $A + B + A - B -$ repeat. The sequence is an event-based single-cycle operation. At the end of any cycle the cylinders must all be in the start position. In this case all the cylinders must be fully retracted at the end of the cycle. In any complete sequence there must be the same number of extend operations ($+$ signs) as retract operations ($-$ signs) for each cylinder.

Step 1. Using double-pilot-operated valves the power circuit can be drawn as shown in Fig. 5.25. Indicate on each 5/2 valve which pilot port will extend the cylinder ($+$ port) and which port will retract the cylinder ($-$ port).

 Note that the valves are shown in the at-rest condition, i.e. the condition at the start of the sequence.

Step 2. Obtain the first operation. This is a signal to give $A+$. As it is a single-cycle sequence, i.e. not continuous, a push-button 3/2 start valve will be used to give a ' $+$ ' signal as shown in Fig. 5.26.

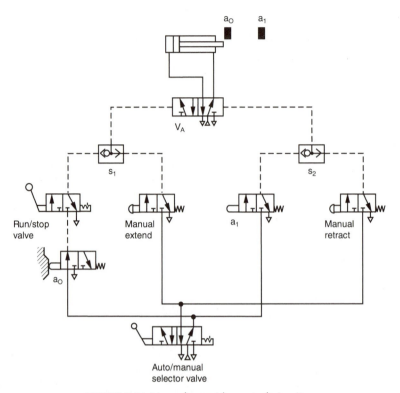

FIGURE 5.24 Manual override control circuit.

Step 3. When cylinder A has fully extended it will operate trip valve a_1 at the end of the stroke. Draw a_1 in the circuit in the unoperated condition (cylinder A is shown retracted). The signal from a_1 is sent to the extend (+) port of the 5/2 valve controlling cylinder B.

Step 4. When cylinder B has fully extended it will operate trip valve b_1 at the end of the stroke. Draw b_1 in the circuit in the unoperated condition (cylinder B is shown retracted). The signal from b_1 is sent to the retract (−) port of the 5/2 valve controlling cylinder A (see Fig. 5.28).

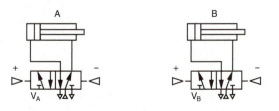

FIGURE 5.25 Step 1: cylinder and power valve arrangement.

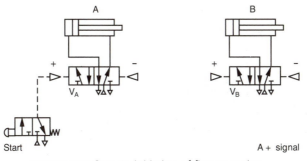

Start   A + signal

FIGURE 5.26 Step 2: initiation of first operation.

Step 5. When cylinder A retracts it will operate trip valve a_0. The output signal from a_0 is used to initiate the next step in the sequence, which is $B-$ as shown in Fig. 5.29.

Note that the trip valve a_0 is shown in the operated condition as cylinder A is retracted in the at-rest condition.

Step 6. The final step in the sequence is achieved when cylinder B retracts.

In order for the circuit to continuously cycle, the last step in the sequence must be used to initiate the first step in the next cycle. Thus the output from the trip valve b_0, which is operated when cylinder B is fully retracted, is fed to the 3/2 stop/run valve which, when switched to the run position, will pass the signal to give $A+$. The circuit for this is shown in Fig. 5.30. It must be noted that the start valve has been changed to a detented stop/run valve.

Consider the same sequence using pilot-operated spring reset 5/2 valves to control the cylinders. These valves are monostable and the signal to the valve has to be maintained to keep the valve in the operated condition. This can be accomplished by using 3/2 trip valves with pilot resets to generate the extend signals, as shown in Fig. 5.31.

To draw the circuit first draw the cylinders in their start position, (–) in this case, both cylinders are retracted. Next draw the cylinder control valves which must pressurise the cylinders into the start condition. The cylinder trip valves a_0, a_1, b_0, b_1, can now be

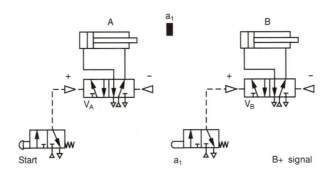

Start a_1 B+ signal

FIGURE 5.27 Step 3: initiation of second operation.

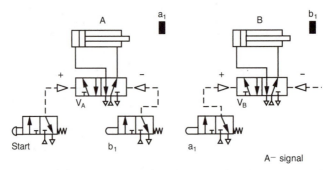

FIGURE 5.28 Step 4: initiation of third operation.

drawn. Do not show the method of resetting the valves yet. (The trip valves generating the A+ and B+ signals will be air reset valves; the other trips will be spring reset valves.) Now draw the stop/run valve.

The last operation, i.e. B−, will be used to initiate the first step A+. Take the output signal from trip valve b_0 (which will be air reset) to give A+ via the stop/run valve. Cylinder A will extend to operate trip valve a_1, which generates the next step, i.e. B+, so the a_1 trip will be air reset. Cylinder B extends operating trip valve b_1, which has to generate the A− signal. This is done by switching off the A+ signal, i.e. resetting valve b_0, and trip valve b_1 will be spring reset. Cylinder A now retracts, operating a_0 which in turn resets a_1, switching off the B+ signal allowing the spring to reset the 5/2 valve retracting cylinder B. With B retracted trip valve b_0 is operated, and this will start the cycle again provided the stop/run valve is in the run position.

5.4.3 Trapped signals in multi-cylinder sequences

Consider the sequence A+ B+ B− A−. Start to draw this sequence using double-pilot-operated valves to control the cylinders, as shown in Fig. 5.32.

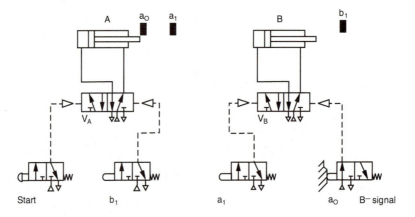

FIGURE 5.29 Step 5: initiation of final operation.

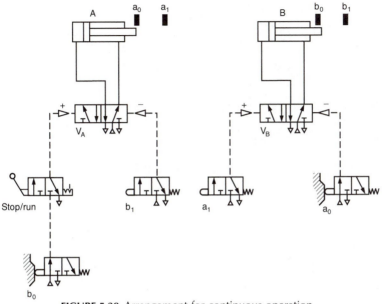

FIGURE 5.30 Arrangement for continuous operation.

Operating the stop/start valve will give an A+ signal, and cylinder A extends operating trip valve a_1 giving a B+ signal. Cylinder B extends operating trip valve b_1, which generates a signal for the next step in the sequence which is B−. As cylinders A and B are both now extended, both trip valves a_1 and b_1 are operating, sending signals to both pilots of valve V_B. There is now a trapped signal on the '+' side of V_B and so the valve will not move to the '−' condition.

To enable the circuit to operate the trapped signal V_B+ must be eliminated by some method. Signal breakers were discussed earlier in the chapter. The circuit in Fig. 5.33 shows the use of signal breakers.

5.4.4 Signal–event charts

In event-based pneumatic sequences the completion of one event is used to signal the start of the next. So a chart relating each signal in a sequence to the cylinder event causing it can be used to show if and when trapped signals occur. The signal–event chart for the continuous sequence previously considered (A+ B+ B− A−) is shown in Fig. 5.34.

The A+ signal is generated when cylinder A is fully retracted and is switched off as soon as cylinder A starts to extend; it only occurs for a short period of time. Similarly for the B− signal. From inspection of the chart it can be seen that the A+ signal is generated while the A− signal exists, as do the B− and B+ signals. For the sequence to operate some form of signal breaker has to be employed to switch off the A− signal before the A+ signal occurs, and another signal breaker to switch off the B+ signal before the B− is generated. Whenever a pair of trapped signals occurs it is the first signal generated that has to be switched off before the second signal can function.

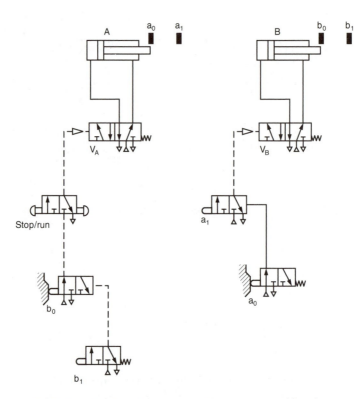

FIGURE 5.31 A+ B+ A− B− circuit using monostable valves.

5.4.5 Trapped signals and spring return cylinder valves

When using pilot-operated spring reset cylinder control valves for the sequence A+ B+ B− A−, trapped signals are again encountered. As the cylinder control valves are

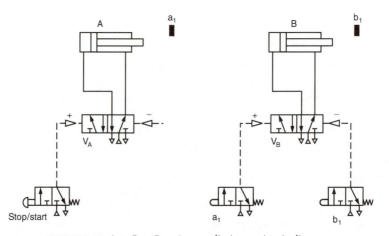

FIGURE 5.32 A+ B+ B− A− preliminary circuit diagram.

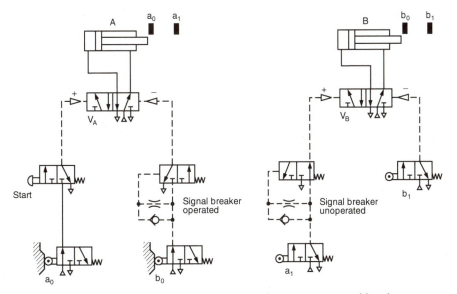

FIGURE 5.33 A+ B+ B− A− circuit diagram using signal breakers.

monostable the signal to them has to be maintained; thus signal breakers cannot be used and other ways must be found. One such method is to use 5/2 trip valves to switch the pilot supply on and off, and such a circuit is shown in Fig. 5.35.

Follow the circuit through starting with the stop/run valve giving A+. This causes trip valve a_1 to operate supplying pilot air through valve b_1 to give B+. Cylinder B extends to operate valve b_1 and release valve b_0. Valve b_1 switches off the pilot signal to

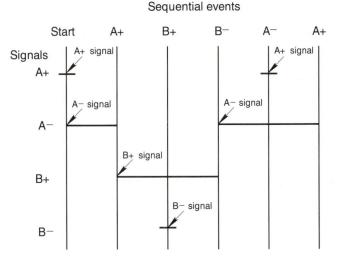

FIGURE 5.34 Signal–event chart for the sequence A+ B+ B− A−.

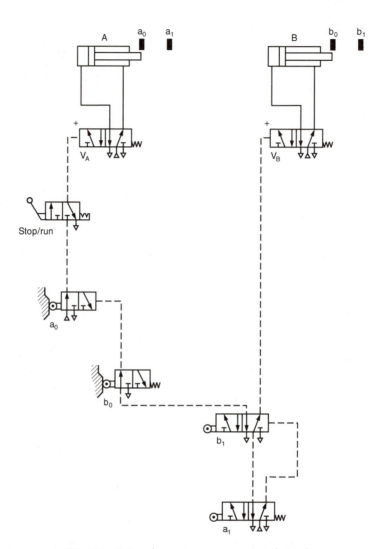

FIGURE 5.35 Trip valves to overcome trapped signals.

B+ and at the same time supplies pilot air to valve b_0. Cylinder B retracts, valve b_1 remains operated and valve b_0 is operated which resets valve a_0, switching off the A+ pilot. Cylinder A retracts and valve a_1 is released. This sends a pilot signal to reset valve b_1 which removes the reset signal from valve a_0. When cylinder A is fully retracted valve a_0 is operated sending an A+ pilot signal to the stop/run valve. This is quite a complex circuit and is generally only used for special applications when spring reset valves form a feature of the system, possibly for automatic retraction or extension of a cylinder if there is a failure of the pilot air.

Example 5.2

A pneumatically operated machine is shown diagrammatically in Fig. 5.36. It is to feed a component from the magazine and then clamp it. As a component is fed to the clamp it ejects the previous one. The sequence of operations is to be:
feed component, clamp, retract feed cylinder and then unclamp. The sequence is to be continuous provided there are components in the magazine.

    As a safety feature the clamp cylinder must be powered forward in the event of a failure in the air supply.

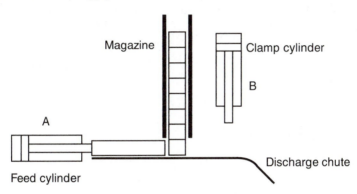

Sequence A+ B+ A− B−

FIGURE 5.36 Diagrammatic representation of feed and clamp machine.

Solution

The sequence will be:

$$A+ \ B+ \ A- \ B-$$

As the clamp cylinder has to clamp the component and not extend to a fixed position, pressure sensing must therefore be used to detect the B+ condition. The machine must stop when the magazine is empty and a valve must be incorporated in the circuit to stop the cycle when this occurs.

    In the event of air supply failure the clamp cylinders must be energised. This can be achieved by using a spring offset cylinder control valve and an air reservoir.

    Figure 5.37 shows a possible circuit for this machine. The sequence is initiated by a signal from trip valve b_0 passing through the magazine trip valve, provided there is a component present, onto the stop run valve to give an A+ signal. Cylinder A extends and operates trip a_1 which is an air reset valve. When this trip is operated it switches off the B− signal to the spring offset valve V_B, which allows the spring to reset the valve, extending the clamp cylinder B. When cylinder B is fully clamped there will be zero air pressure in the annulus side and on the pilot of the diaphragm valve b_1, which resets under the action of its spring, sending an A− signal to valve V_A. Cylinder A retracts, operating trip valve a_0 which sends a signal

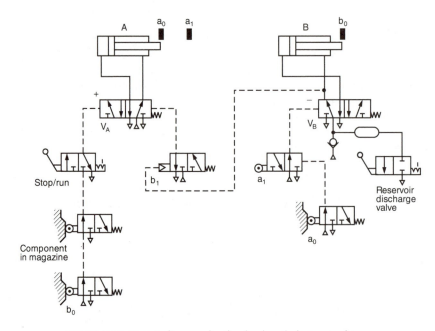

FIGURE 5.37 Circuit diagram for the feed and clamp machine.

to reset trip valve a_1 which, in turn, gives a B− signal. Cylinder B retracts, operating trip valve b_0 and the sequence is ready to restart.

Should there be a failure of the air supply, valve V_B will be reset by its spring causing air from the reservoir to flow to the full bore side of cylinder B energising the clamp. The check valve by the air reservoir stops the air from the reservoir discharging if the supply is at fault. It is advisable to fit a shut-off valve to the reservoir to allow it to be discharged when required.

5.4.6 Two steps initiated from one signal

In some sequences it may be desirable for two cylinders to start moving at the same instant. This is often done to reduce cycle time. Consider the sequence

$$A+ B+ A-$$
$$B-$$

The notation used indicates that both cylinder A and cylinder B are signalled to start retracting at the same time. This does not necessarily mean that both cylinders actually start to retract simultaneously or reach the fully retracted state together. The circuit in Fig. 5.38 uses double-pilot-operated 5/2 cylinder valves.

In the circuit in Fig. 5.38, the trip valve b_1 sends a retract signal to valves V_A and V_B at the same time. There is an extend signal already applied to V_B from limit switch a_1, this prevents valve V_B moving across until the ' + ' signal is removed: thus cylinder A will start to retract first, and cylinder B will retract when trip valve a_0 is reset by its spring. To ensure that both cylinders A and B are fully retracted before the cycle can

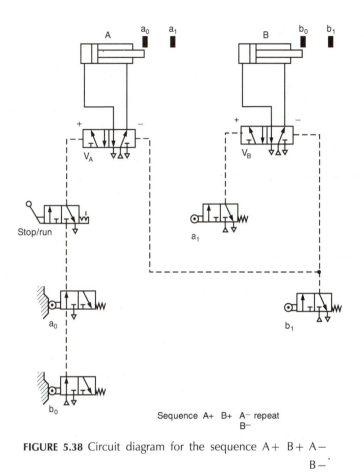

FIGURE 5.38 Circuit diagram for the sequence A+ B+ A−
                                                 B−

recommence, trip valves a_0 and b_0 are connected in series acting as a logic AND gate. Both trip valves have to be operated before a signal is sent to the stop/run valve.

5.5 Cascade circuits

The cascade principle has been developed for pneumatic sequences where trapped signals may occur. The sequence is divided into a number of sections or groups which do not contain a trapped signal. The pilot air is switched so that only one group is supplied with pilot air at any one time. This is done using a series of cascade valves.

5.5.1 Designing a cascade circuit

To design a cascade circuit a series of steps are followed, as illustrated in the following example.

Consider the sequence A+ B+ B− A− using double pilot-operated direction control valves.

1. Divide the sequence into groups so that no letter repeats in any group.

$$A+ \ B+ \ \Big/ \ B- \ A-$$
$$\text{Group I} \ \Big/ \ \text{Group II}$$

2. Draw the cylinders together with their associated control valves and trip valves. Add a stop/run valve.
3. Draw in pilot air changeover valves (cascade valves) at the bottom of the circuit. If 5-port valves are used, one less valve than the number of groups is needed. In this example only one cascade valve is required as there are only two groups.
4. Draw in group lines I and II above the cascade valves with pilot air on group I. The group lines are sometimes referred to as 'bus bars'. Where possible use different coloured lines for the group lines; in this example chain-dotted and long-dotted lines have been used.
5. The sequence starts with $A+$, which is the first step in group I. So, to initiate the sequence take a signal from group I 'bus bar' and feed it through the stop/run valve to V_A+.

   In general, if the sequence does not start with a new group, for example,

$$A+ \ \big| \ A- \ B+ \ \big| \ B-$$
$$\text{I} \ \big| \ \text{II} \ \big| \ \text{I}$$

take a signal in the appropriate group from the stop/run valve to initiate the first step. Work through the sequence and use the last operation to supply a signal to the stop/run valve which will start the sequence over again.

The steps 2 to 5 are shown in Fig. 5.39.

6. Cylinder A extends ($A+$). This is in group I, so supply trip valve a_1 with group I pilot air. Take the output to valve V_B to give the next step $B+$.
7. Cylinder B will extend. This operation is still in group I, so supply trip valve b_1 with group I pilot air. The output signal from b_1 is used to switch the cascade valve which exhausts group I lines and supplies air to the group II lines.
8. The first step in group II is initiated by feeding group II pilot air directly to the '$-$' side of valve V_B causing cylinder B to retract ($B-$).
9. When cylinder B retracts, it operates b_0 which is supplied with group II air. The output from b_0 trip is fed to the '$-$' side of valve V_A to give $A-$.
10. Cylinder A retracting operates trip valve a_0 which again is supplied with group II air. The output from a_0 is used to switch the group change valve back to group I.
11. The cylinders have now been through a complete cycle. The last operation in the cycle is used to initiate the next, in this case the change to group I, with group I pilot air being fed to the stop/run valve.

The cycle is now complete, as illustrated in Fig. 5.40.

5.5.2 Grouping a sequence

When splitting a sequence into groups for a cascade circuit ensure that the minimum number of groups is determined. Split the sequence into groups starting at the beginning

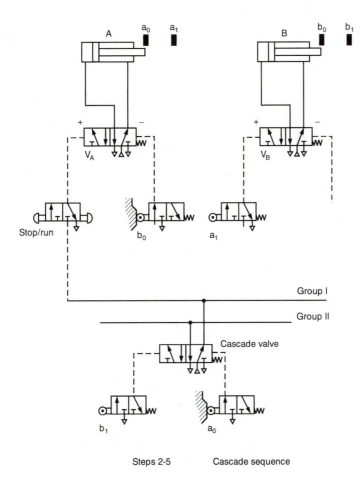

Steps 2-5   Cascade sequence

FIGURE 5.39 Steps 2 to 5 in the construction of a cascade circuit.

or end of the sequence. This is forward and reverse splitting. Consider the following sequence shown split in both forward and reverse directions:

Forward splitting   A+ B+ | A− C+ D+ | D− C− B−

                 I   II   III

Reverse splitting   A+ | B+ A− C+ D+ | D− C− B−

                 I   II   I

By reverse splitting, the latter part of the sequence joins onto the start and only two groups are required.

5.5.3 Rules for the design of a cascade circuit

1. Divide sequence into groups so that no trapped signals can occur, i.e. no letter repeats in a group.

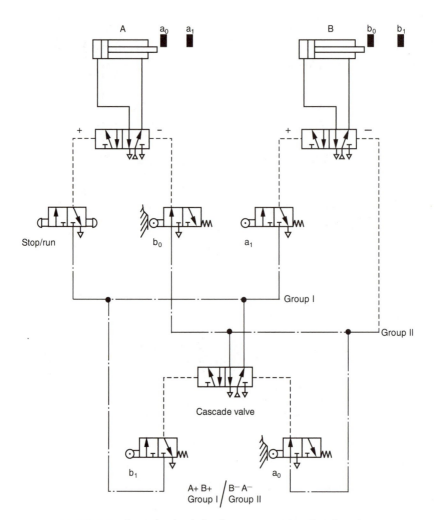

FIGURE 5.40 Cascade circuit for the sequence $A+ B+ B- A-$.

2. Draw the circuit components, cylinders, direction control, trip and stop run valves. Mains air can be shown supplied through the direction control valves to set the cylinders into their start conditions.
3. Draw the cascade valves and group pilot air lines.
4. All the pilot valves in a group are supplied with air from that group. The signal from the last step in a group is used to switch the appropriate cascade valve and thus step the pilot air onto the next group.

The first operation in a group is initiated by pilot air directly from the group air line. With other methods of overcoming trapped signals, such as signal breakers, switch valves or pulse valves, it is necessary to identify exactly where the trapped signal will occur.

Cascade circuits avoid this problem because – providing the above procedure and rules have been obeyed – air will be supplied to a particular section of the control circuit only when it is needed. Although trip valves triggered earlier in the sequence may remain in the operated condition, no signal is transmitted because their air supply has been removed.

Cascade circuits offer a relatively foolproof system for the control of many sequences. This does not mean that they necessarily present the best solution in every situation. They can, however, provide a sound basis for developing a more practical or economical circuit.

5.5.4 Use of signal–event charts

Consider the sequence A+ B+ C+ A− B− C−. Dividing this sequence into groups using the cascade method will result in two groups:

$$A+ \ B+ \ C+ \ \bigg| \ A- \ B- \ C-$$
$$\quad\ \ \text{I} \qquad\qquad \text{II}$$

Construct the signal–event chart for this sequence (Fig. 5.41).

At the start of the sequence the step C− generates the A+ signal which remains on until cylinder C starts to extend, i.e. until C+. Similarly, the A+ step generates the

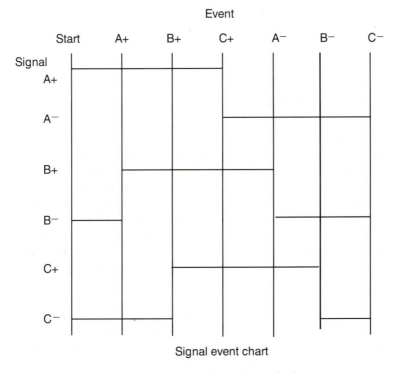

Signal event chart

FIGURE 5.41 Signal–event chart for a three-cylinder circuit.

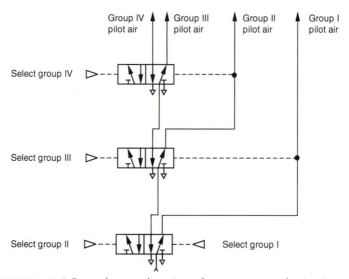

FIGURE 5.42 5-Port valves used to give a four-group cascade circuit.

next step, i.e. B+, and so on. The completed chart (Fig. 5.34) shows that at no stage in the sequence does a trapped signal occur and so the sequence can be achieved without the use of a cascade circuit.

5.5.5 Cascade valve arrangements

Figure 5.42 shows the 'Cascade' for a four-group sequence using 5/2 valves.

In the diagram, group I is activated; all the other groups are exhausted. The output signal from the last trip valve in group I is used to switch the bottom cascade valve, which puts air into group II, at the same time exhausting group I. Group II air is used to reset the top cascade valve so it is in the correct position ready for the next group change. The signal from the last trip in group II is used to switch the middle cascade valve which puts air into group III, exhausting group II. The last signal from group III is used to switch the top valve to group IV and the last signal from group IV resets the bottom valve, putting air back into group I. Air in the group I line resets the middle valve ensuring that it is ready for the next change. At any particular time there is only one group line pressurised all the other lines are exhausting.

In the cascade system shown in Fig. 5.42 the pilot air to group IV has to flow through all three cascade valves before going to the trip valves. This can give rise to considerable pressure drops and will result in slow responses and even malfunction of the circuit. A method of overcoming this difficulty is to provide one cascade valve per group. Depending upon the system adopted, the cascade valves may be 3 port or 5 port and there are advantages and disadvantages associated with each.

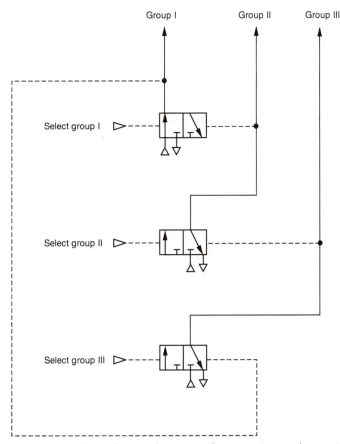

FIGURE 5.43 3-Port valves used to give a three-group cascade circuit.

5.5.6 Three-port valve cascade system

Figure 5.43 shows one 3/2 valve used for each group. On a group change, pilot air from the new group is used to switch off the preceding group. Thus one group line can be pressurised before the preceding group line is exhausted. In the event of a cascade valve failing to reset, it is possible to have pilot air pressure supplied to two groups at the same time, which could be dangerous. Problems particularly arise with fast cycling systems.

5.5.7 Alternative 5-port valve cascade system

In this arrangement (Fig. 5.44) there are as many 5-port cascade valves as there are groups, but the air supply to each group has to pass through two cascade valves. On a group change, disconnection of the existing pilot supply and connection of the next occurs simultaneously reducing the possibility of having two groups pressurised at the same time.

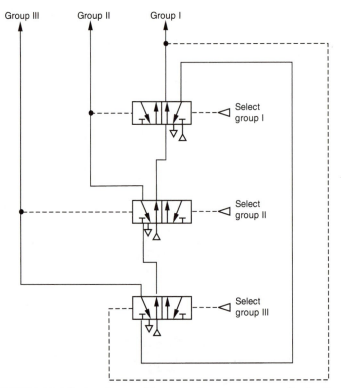

FIGURE 5.44 5-Port vavles used to give a three-group cascade circuit.

It should be noted that the valves are connected in a non-standard manner with mains air switched through ports 3 and 5 and port 1 functioning as an exhaust port. Care must be taken in choosing the valves. Most spool-type valves are suitable for this method of operation but many poppet and slide valves are not.

5.5.8 Modified cascade circuits

Cascade circuits are primarily intended for systems which use trip valves to detect completion of cylinder movements. Where other methods of sensing are to be included the cascade principle can still be used to create the initial design for a sequence. The basic circuit can then be modified, incorporating pressure sensing, time delays, logic function and so on. The cascade method can also be used with solenoid-operated valves controlled by relay logic or programmable logic controllers (PLCs).

Example 5.3

A machine which is magazine fed, drills and taps a hole in a flat component. A schematic layout is shown in Fig. 5.45.

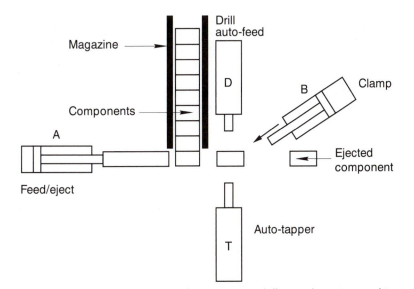

FIGURE 5.45 General arrangement of an automatic drilling and tapping machine.

The feed cylinder A pushes a component forward until A is fully extended, and this action automatically ejects the previous workpiece. The component is now clamped in position by cylinder B. The self-feed drill is next energised; this unit automatically extends a pre-set distance and then retracts. When fully retracted it sends a pneumatic pulse to start an automatic tapper which is located beneath the workpiece in line with the drill unit. The auto-tapper operates in a similar way to the self-feed drill: it extends, tapping to a pre-set depth, and then automatically reverses, screwing itself out until it is fully retracted, at which stage it sends a pneumatic pulse for the next operation.

Let the auto-feed drill (AFD) be D and let the auto-feed tapper (AFT) be T. Essential steps in the sequence will then be

$$A+ \ B+ \ D+ \ D- \ T+ \ T- \ B- \quad \text{Repeat}$$

The A− signal can occur any time after the B+ operation. The D+ and D− signals can be considered as one signal D as the D− signal is automatically generated by the drill unit when it has completed its cycle; this also applies to the self-feed tapper. The sequence now becomes

$$A+ \ B+ \ \left| \begin{array}{l} D \ T \ B- \\ A- \end{array} \right.$$

Group I   Group II

Assuming the retract signal on the feed cylinder occurs after the clamp operates, a two-group cascade can be devised. Because correct clamping relies on the clamping pressure and not on the position of the cylinder rod, the extend operation of the clamp cylinder must be pressure sensed.

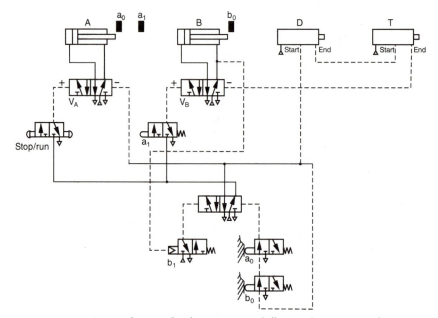

FIGURE 5.46 Circuit diagram for the automatic drilling and tapping machine.

The completion signal from the AFD initiates the AFT which, on ending its cycle, gives a signal retracting the clamp cylinder. The clamp retract signal is interlocked with the feed cylinder retract trip valve to ensure that both cylinders are fully retracted before the next cycle can be initiated. The resulting circuit is shown in Fig. 5.46.

Note: The a_0 trip and b_0 trip act as a logic AND gate; this ensures that cylinder A is fully retracted before the next cycle starts.

5.5.9 Cascade circuits using spring offset valves

The cascade circuits so far described have all used double-piloted main valves to control the cylinders. The use of spring offset valves can be advantageous in certain applications – for example, where a cylinder must take up a safe position in the case of mains air failure. This can be achieved using an air receiver and a spring offset valve. An example of this was shown in Fig. 5.13 earlier in this chapter. The air receiver has to be of sufficient capacity to fully extend the clamp cylinder B and give an adequate clamping force.

Example 5.4

A pneumatically operated pick and place unit shown schematically in Fig. 5.47 is used to move components from point 1 to point 2 on a conveyor.

The sequence of events must be: Position gripper at point 1, grip component, lift component, move to above point 2, lower, release component, lift and return to start position.

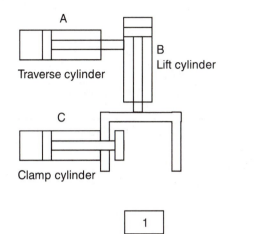

FIGURE 5.47 General arrangement of a pick and place machine.

This can be expressed as a sequence, commencing with the three cylinders retracted as

$$B+ \ C+ \ B- \ A+ \ B+ \ C- \ B- \ A-$$

It will be noted that cylinder B operates twice in the sequence, so it may be necessary to have a pair of trip valves or sensors at each end of the stroke.

From safety considerations the component must remain gripped if there is a failure of the air supply. This can be done by using an air receiver similar to that shown in Fig. 5.13, or by using a spring-loaded clamp or clamp cylinder which has to be powered open. In this example a spring-loaded clamp will be used.

Consider next how the circuit may be simplified or the sequence time reduced. Can two of the steps occur at the same time? Can any steps be eliminated? Can the steps occur in any other sequence? In this instance all the steps are essential and the sequence cannot be altered; however, cylinder A can extend as soon as cylinder B starts to lift, and, similarly, A can start to retract as soon as B starts to lift. The sequence can be modified to

$$B+ \ C+ \ B- \quad B+ \ C- \ B-$$
$$\text{delay } A+ \qquad \text{delay } A-$$

One signal can be used to initiate $B-$ and $A+$ with a pneumatic delay to the $A+$ signal.

Dividing the sequence into groups gives

B+ C+		B−	B+ C−		B−
	delay A+			delay A−	
I	II		III	IV	

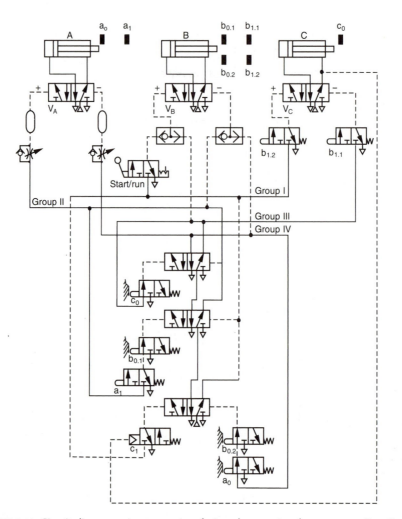

FIGURE 5.48 Circuit diagram using two pairs of trip valves to give the sequence B+ C+ B−
A+ B+ C− B− A−.

As C is a clamping cylinder, pressure sensing has to be used to detect satisfactory
completion of C+. A trip valve can be used to detect C−. The circuit for this
sequence is shown in Fig. 5.48. Time delays are shown on both pilot ports to valves
V_A, and shuttle valves to the pilots of V_B.

The clamping device is to be spring loaded to simplify the operation of the
clamping cylinder. A single-acting cylinder can be used to de-energise the clamp.
The modified circuit for the operation of cylinder C is shown in Fig. 5.49 and uses
a 3/2 valve V_C.

The ' − ' pilot on V_C gives unclamped condition and the ' + ' pilot clamped.

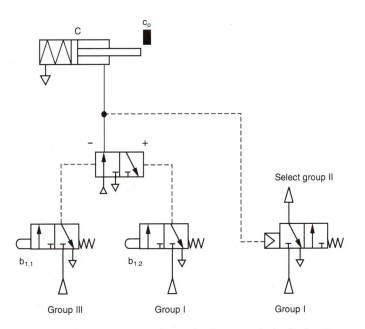

FIGURE 5.49 Alternative circuit design for the control of cylinder C.

5.6 Hydro-pneumatics

Pneumatic systems have two major limitations when compared with hydraulic systems.

1. They are very compressible. So in applications where steady motion, precision speed control or positive positioning are required, a simple pneumatic actuator is unsuitable. An equal pressure hydro-pneumatic system in which the compressed air is used to pressurise a liquid (usually mineral oil), and then the almost incompressible oil used in an actuator, largely overcomes this problem.
2. Compressed air is normally supplied at a pressure of 7 bar gauge or less in most installations. This pressure limits the maximum thrust available from pneumatic cylinders to about 10 kN unless extremely large cylinders are used. An air–oil intensifier powered by a normal compressed air supply can be used to deliver oil at high pressure.

5.6.1 Equal pressure air–oil systems

These are used to give constant speed, smooth slow speeds and positive locking of a cylinder.

A car service lift as used in garages often employs a hydro-pneumatic system, as shown diagrammatically in Fig. 5.50. To raise the load, valve B is operated, pressurising the air–oil reservoir; valve A is then operated to allow the pressurised oil to flow into the cylinder and raise the load. When neither valve A nor valve B is operated the cylinder

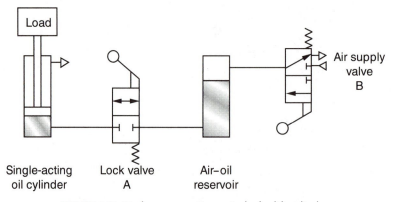

FIGURE 5.50 Hydro-penumatic control of a lift cylinder.

will be hydraulically locked in position; increasing or reducing the load will have virtually no effect on its position. The cylinder can be left locked in position under full load for long periods of time with no change in position of the load. As oil is much more viscous than air, it is far easier to seal in the system than air. Lowering the load is achieved by operating valve A only.

5.6.2 Speed control using air–oil systems

A liquid is almost incompressible; it does not expand when its pressure drops after flowing through a restrictor as does air. It is therefore much simpler to accurately meter the flow of a liquid. The hydro-pneumatic system shown in Fig. 5.51 is used to give precision speed control in both directions. The accuracy of the control achieved depends upon the type of flow control valves. When the cylinder extends, the flow of fluid out of the cylinder is being metered; this is termed 'meter-out' flow control. This creates a back pressure in the cylinder and prevents any tendency for the cylinder load to run away. When the cylinder is retracting, raising the load, the fluid is being metered as it enters the cylinder; this is known as 'meter-in' speed control. As there are two flow control valves and associated check valves used in this circuit, it is possible to obtain independent speed control on extend and retract. If there is no possibility of the load overrunning or running away, and the same speed is required on both extend and retract, a single restrictor without a check valve may be used to make the flow in both directions.

The dynamic thrust obtained from the cylinder is much reduced due to viscous friction of the liquid causing pressure drops within the system. The pressure drops are very difficult to calculate and will vary with the flow rate and temperature of the liquid. It is therefore usual to use the empirical formula:

Dynamic thrust = 0.3 × Static thrust for a hydro-pneumatic system.

If a standard pneumatic cylinder is used with liquid on one side of the piston and air on the other, there is a possibility of air leaking across the piston into the fluid. This will result in an erratic or spongy operation. Commercial hydro-pneumatic cylinders have a special sealing arrangement on the piston to prevent air leaking into the fluid side.

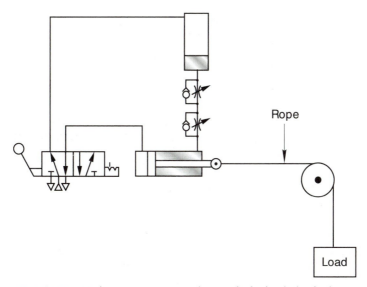

FIGURE 5.51 Hydro-pneumatic speed control of a loaded cylinder.

5.6.3 Double-acting hydro-pneumatic cylinders

The problem of air leakage across the piston seal can be eliminated by using fluid on both sides of the piston. Two reservoirs are connected to the cylinder, as shown in Fig. 5.52, the reservoirs being alternately pressurised to extend and retract the cylinder.

Using the 5/3 direction control valve to supply air to the air–oil reservoirs allows the cylinder to be accurately stopped partway along its stroke. In order to lock the cylinder in any intermediate position along its stroke, lock valves must be fitted between the cylinder and the reservoirs, as shown in Fig. 5.53. When the 5/3 valve is centralised the lock valves are sprung into the closed position and the cylinder is positively locked.

Operating the 5/3 valve to extend or retract the cylinder automatically supplies pilot air to the lock valves via the shuttle valve to allow the cylinder to move.

5.6.4 Emergency stop in an air–oil system

If a cylinder has to be stopped and locked in position in case of an emergency such as a power failure, a hydro-pneumatic circuit can be utilised. Consider the circuit shown in Fig. 5.53. This can be modified by the insertion of a manual stop/run valve, as shown in Fig. 5.54.

When the switch valve V1 is not piloted, the cylinder locking valves are spring operated into the closed position and the cylinder is locked in position. Should there be a failure of the air supply, or if the emergency stop button is operated, the cylinder will immediately lock. If the emergency stop button is remote from the switch valve there will be a delay in the cylinder locking; this could well be unacceptable and in such a case electrical operation may provide a solution, with the switch valve V1 being replaced by a solenoid-operated valve. A possible electrical circuit shown in Fig. 5.55 incorporates

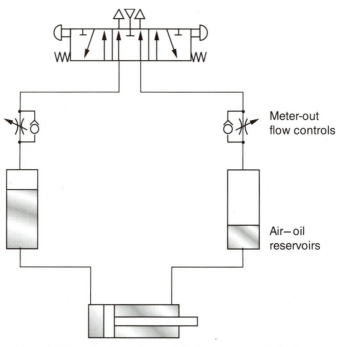

FIGURE 5.52 Using two oil reservoirs to overcome air leakage.

multi-emergency stop switches and a pneumatic pressure switch in case of an air supply failure.

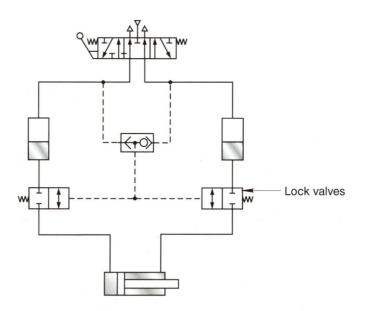

FIGURE 5.53 Hydro-pneumatic control with intermediate locking.

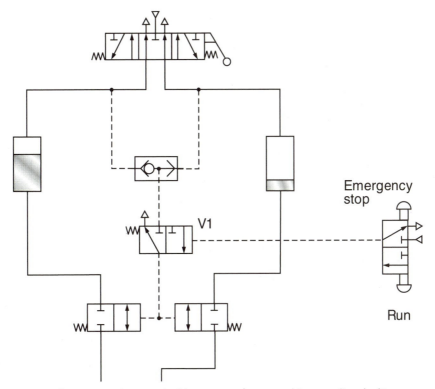

FIGURE 5.54 Hydro-pneumatic control with emergency stop and intermediate locking.

5.6.5 Air–oil reservoirs

There are several types commercially available, the commonest being based on a standard cylinder body. The reservoir must be mounted vertically to prevent air from going into the oil system. It is recommended that the capacity of the reservoir be at least 1.25 times the volume of oil required by the cylinder stroke, i.e. the swept volume. The reservoir length should be at least three times its diameter in order to prevent air leaving the reservoir with the oil. A typical air–oil reservoir is shown diagrammatically in Fig. 5.56.

Other types of air–oil reservoirs separate the fluids by either a bladder or a free piston. This eliminates air entrainment with the oil. It is possible to use a pneumatic cylinder as an air–oil reservoir, the position of the piston rod indicating the volume of oil in the reservoir.

5.6.6 Duplex air–oil cylinders

A duplex cylinder can be used as an air–oil cylinder, the rod end section being used as a through-rod oil cylinder. There may, however, be some slight leakage from the oil side of the cylinder, and to overcome this it is usual to fit a small make-up reservoir which may be either spring loaded or pressurised with air. The two sides of the oil cylinder are

---

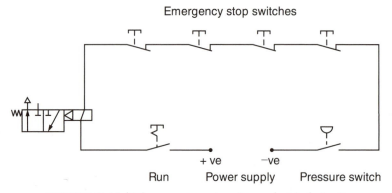

FIGURE 5.55 Multiple emergency stops in an electrical circuit.

connected by flow control valves to give speed control, the other section of the cylinder being connected through a valve to the air supply, as shown in Fig. 5.57.

5.6.7 Hydro-checks

These are commercial units which can be used to damp or control the speed of a linear actuator. They may be inward or outward checking, or may operate in both directions. They consist basically of a cylinder with a spring-loaded make-up reservoir to compensate for the difference between the full bore and annulus volumes of the cylinder. Figure 5.58 shows the basic operating principle. When the hydro-check cylinder is extending, oil flows from the reservoir to the full bore side, and when the cylinder retracts oil flows into the reservoir.

The screwed rod on the hydro-check piston rod enables the piston rod of the cylinder to be controlled and to move freely for part of its stroke prior to the hydro-check coming into operation. This free movement is sometimes called the skip distance. This type of hydro-check may be used to control the movement of a drilling machine, the

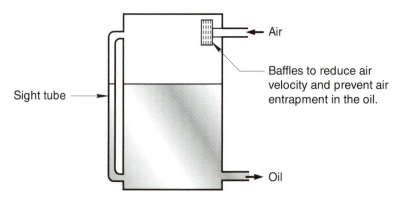

FIGURE 5.6 General arrangement of an air–oil reservoir.

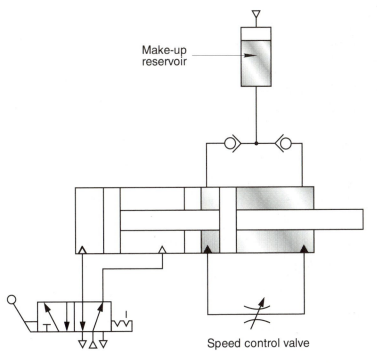

FIGURE 5.57 General arrangement of a duplex air–oil cylinder.

skip distance giving a rapid approach of the drill bit to the workpiece, the hydro-check speed control coming into operation just before the drill contacts the workpiece. In this application the retraction of the drill may be at full speed throughout the whole of its stroke.

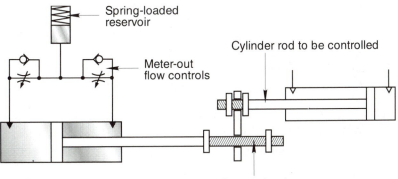

FIGURE 5.58 General arrangement of a hydro-check with skip facility.

Example 5.5

A machine table has to be reciprocated through a distance of 750 mm. The extend speed is to be adjustable between 20 and 100 mm/min, the forward thrust required has a maximum value of 1.3 kN. The return stroke is to be at approximately 1 m/min. Design a suitable hydro-pneumatic circuit sizing the cylinder and any air–oil reservoirs used. The air supply available has a maximum pressure of 6 bar gauge.

Solution

As the extend speed is very low a hydraulic control is to be used. The retract speed is comparatively fast and so a pneumatic drive can be used.

In a hydro-pneumatic system

$$\text{Dynamic thrust} = 0.3 \times \text{Static thrust}$$

Let the cylinder diameter be D

$$\text{Static cylinder thrust} = \text{Pressure} \times \text{Area}$$

In this case

$$\text{Pressure} = 6 \text{ bar gauge}$$

$$= 6 \times 10^5 \text{ N m}^{-2}$$
$$\text{Dynamic thrust} = 1.3 \text{ kN} = 1300 \text{ N}$$

Thus

$$\text{Dynamic thrust} = 0.3 \times \text{Static thrust}$$

$$= 0.3 \times 6 \times 10^5 \times \frac{\pi D^2}{4}$$

Therefore

$$D^2 = \frac{1300 \times 4}{0.3 \times 6 \times 10^5 \times \pi} \text{ m}^2$$

$$D = 95.6 \text{ mm}$$

The nearest standard metric cylinder greater than the calculated value is 100 mm bore. As this will give a slightly higher thrust than required, a pressure regulator could be used to reduce the air supply pressure. The circuit would then be as shown in Fig. 5.59.

The volume of oil stored in the reservoir should be 25 per cent more than that required by the cylinder.

The maximum volume of oil in the cylinder is the area of the piston times the stroke. Hence,

$$\text{Volume of reservoir} = 1.25 \times \frac{\pi D^2}{4} \times L$$

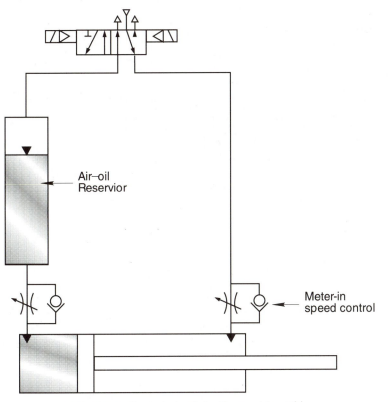

FIGURE 5.59 Control circuit for the machine table.

where $D = 0.1$ m and $L = 0.75$ m. Thus,

$$\text{Reservoir volume} = 1.25 \times \pi \times \frac{0.1^2}{4} \times 0.75$$

$$= 0.00736 \text{ m}^3$$

$$= 7.36 \text{ litres}$$

If the reservoir is to be manufactured, its length should be three times its diameter as already explained. Free surface reservoirs must be mounted vertically.

5.7 High-pressure air–oil systems

Hydraulic fluid is raised to a pressure greater than that of the air supply by using an intensifier or an air-driven hydraulic pump. The high-pressure hydraulic fluid can be fed directly to a hydraulic actuator or controlled by hydraulic valves.

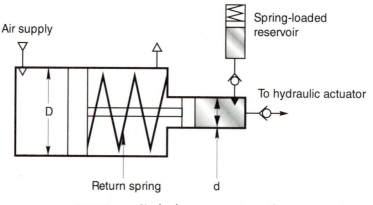

FIGURE 5.60 Single-shot pressure intensifier.

5.7.1 Single-shot intensifiers

These consist of a large-bore pneumatic cylinder driving a small-bore hydraulic cylinder, shown diagrammically in Fig. 5.60. The small oil reservoir is used to compensate for any leakage on the hydraulic side of the system.

The intensification ratio is equal to the square of the ratio of the cylinder diameters.

$$\text{Intensification ratio} = \frac{\text{Hydraulic pressure}}{\text{Pneumatic pressure}}$$

$$= \frac{D^2}{d^2}$$

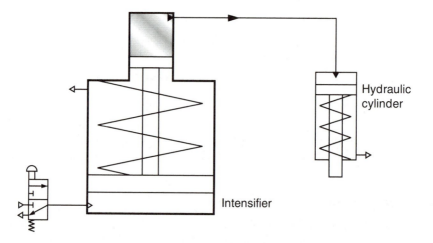

FIGURE 5.61 Simple application of an air–oil intensifier.

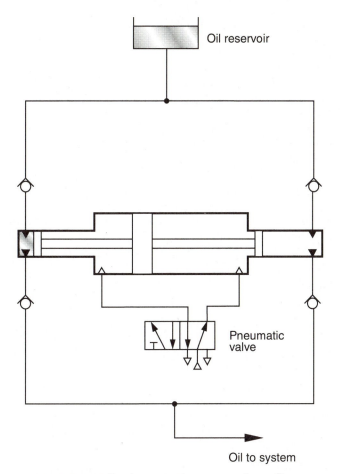

FIGURE 5.62 Continuous action pressure intensifier.

Commercial single-shot intensifiers are generally available for intensification ratios from 10:1 to greater than 100:1, different ratios being obtained by using a standard size pneumatic cylinder with a different bore hydraulic cylinder. Thus, as the pressure ratio increases, the quantity of fluid delivered per stroke reduces.

This type of intensifier is normally used to power short-stroke single-acting hydraulic cylinders for clamping, marking, bending, piercing, etc. The intensifier manufacturers produce a range of suitable cylinders with thrusts of up to 200 kN but with relatively short strokes in the order of 15 mm. An air–oil intensifier can be incorporated into a pneumatic system to give high forces more economically than can be produced from a pure pneumatic system. A simple layout is shown in Fig. 5.61.

The return speed of the hydraulic cylinder is limited by the rate at which air is exhausted from the intensifier, although this rate can be increased by fitting a quick exhaust valve. The hose and fittings connecting the intensifier to the hydraulic cylinder must be capable of withstanding the maximum oil pressure that the intensifier can develop.

The output pressure of the intensifier can be controlled by using a pressure regulator on the input air supply.

5.7.2 Continuous pressure intensifiers

These are similar to the single-shot intensifiers but the pneumatic cylinder is continually cycled when it is supplied with compressed air. A diagrammatic representation of a continuous intensifier is shown in Fig. 5.62.

Continuously reciprocating air–oil intensifiers are commercially available with operating pressures up to and in excess of 700 bar. The quantity of oil that these units can deliver is limited by the oil reservoir capacity, the flow rates being in the order of up to 0.2 l/min. The units only consume air when delivering oil and stall once the set oil pressure has been reached. They can be used to power low flow rate hydraulic circuits.

Some pneumatic valves can be used with lower pressure hydraulic fluid, e.g. manually or mechanically directly operated direction control valves with threaded exhaust ports may be used at pressures of up to about 20 bar. However, pilot-operated solenoid-controlled valves must not be used with a hydraulic fluid.

Example 5.6

A table which moves vertically is used to stack sheets of material whose thickness is pre-set at any value between 10 and 50 mm. The table is arranged so that, at the top of its travel, the first sheet can slide onto it and the table is then lowered by an amount equal to the thickness of the sheet. A second sheet can then be fed onto the table, and this process is repeated until the table is at the bottom of its travel. The table is powered by a hydraulic cylinder and is arranged generally as shown in Fig. 5.63.

Maximum production rate is 4 sheets per minute. Assume that it takes 10 seconds to lower the table and 15 seconds to slide the sheet onto the table. The table must be locked at each position. When the table is fully loaded it is raised and the stack of sheets removed.

Design a suitable hydro-pneumatic circuit and estimate the size of a hydraulic cylinder taking the rod area as half the full bore area. A continuous hydro-pneumatic intensifier with a 10 : 1 ratio and a maximum oil delivery of 0.2 l/min is to be used. Neglect all losses and assume a maximum air supply pressure of 5 bar gauge. Hydraulic cylinders are available in 10 mm bore increments.

Solution

With a 10 : 1 intensifier the maximum hydraulic pressure is:

$$10 \times 5 = 50 \text{ bar gauge}$$

Determine the cylinder diameter

$$F = P \times A$$

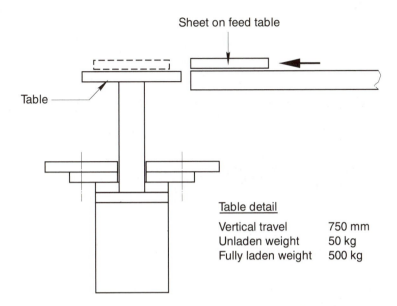

FIGURE 5.63 General arrangement of the sheet-stacking table.

$$D = \sqrt{\frac{4F}{\pi P}}$$

$$D = \sqrt{\frac{4 \times 500 \times 9.81}{\pi \times 50 \times 10^5}}$$

$$D = 35.4 \text{ mm}$$

The nearest cylinder bore is 40 mm. The rod area is to be half the full bore area, so therefore the rod diameter will be

$$\frac{40^2}{2} = 28 \text{ mm}$$

The maximum flow into the cylinder when lowering – the maximum step is 50 mm – can be determined as follows.

Time taken to lower is 10 seconds.

$$\text{Annulus area} \times \text{Velocity}$$

$$= \pi \left(\frac{40^2 - 28^2}{4} \right) \times 10^{-6} \times \frac{50}{10} \times 10^{-3}$$

$$= \frac{\pi(1600 - 784)}{4} \times \frac{50}{10} \times 10^{-9} \text{ m}^3/\text{s}$$

$$= 3200 \times 10^{-9} \text{ m}^3/\text{s}$$

$$= 3200 \times 10^3 \times 60 \times 10^{-9} \text{ l/min}$$

$$= 0.19 \text{ m}^3/\text{min}$$

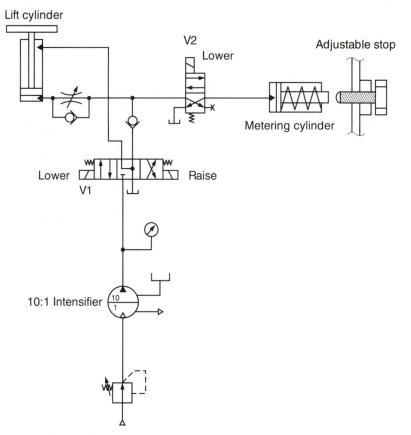

FIGURE 5.64 Hydro-pneumatic circuit to operate sheet-stacking table.

The proposed intensifier with a delivery of 0.2 l/min is adequate.
    The time taken to raise the table is given by

$$\text{Time taken} = \frac{\text{Full bore volume}}{\text{Flow rate}}$$

$$= \frac{\pi \times 40^2 \times 10^{-6} \times 750 \times 10^{-3}}{4 \times 0.2 \times 10^{-3}}$$

$$= 4.7 \text{ min}$$

If this is too slow a regenerative circuit may be used. Full details of regenerative circuits may be found in *Power Hydraulics* by Pinches and Ashby, published by Prentice Hall, together with information on the hydraulics valve used in the circuit for this machine (see Fig. 5.64).
    Consider the circuit shown in Fig. 5.64 with the valve V1 in mid-position. The intensifier output is blocked and will stall off; the full bore side of the cylinder will

be connected through valve V2 to a blocked port. The lift cylinder will be hydraulically locked in the position by the check valve and valve V2.

The metering cylinder will be fully retracted by the action of the spring which may be either internal to the cylinder or external to it. A drain port is fitted to the annulus end to allow any leakage of oil past the piston seal to return to the tank.

Energising the lower solenoids on valves V1 and V2 connects the intensifier's output to the annulus side of the lift cylinder while the full bore side is connected to the metering cylinder through the flow control valve. Oil flows from the lift cylinder into the metering cylinder, the stroke of which can be adjusted by a movable stop. This will regulate the exact distance the lift cylinder lowers. De-energising the lower solenoids locks the lift cylinder in place and resets the metering cylinder. The procedure is repeated until the lift cylinder is at the bottom of its stroke.

The raise solenoid is now energised causing the lift cylinder to be raised. The solenoids can be operated from electrical limit switches tripped by the sheets being fed onto the table or by timers, depending upon details of the sheet production.

CHAPTER 6

Logic

The simple dictionary definition of logic is the art of reasoning. In control engineering a logic system is one in which events occur in a reasoned sequence. In other words, an event will be initiated when a predetermined number of inputs occur.

Consider the occasions when a man would use an umbrella:

(1) when outside and raining and not very windy (if it was very windy the umbrella would blow inside out); or
(2) when outside and snowing and not very windy; or
(3) when outside and hot sunshine (as a sunshade) and not very windy.

Let the various conditions be represented by letters:

$$U = \text{Use umbrella}$$

$$O = \text{Outside}$$

$$R = \text{Rain}$$
$$S = \text{Snowing}$$
$$W = \text{Very windy}$$
$$H = \text{Hot sunshine}$$

An equation can be written stating when the umbrella will be used:

$$U = O \text{ and } R \text{ and not } W$$

$$\text{or}\quad O \text{ and } S \text{ and not } W$$

$$\text{or}\quad O \text{ and } H \text{ and not } W$$

This is a logic equation, the logic command words being:

$$\text{AND,}\quad \text{OR,}\quad \text{NOT}$$

Logic statements or equations can be written down mathematically using Boolean algebra.

6.1 Boolean algebra

This was developed by Professor George Boole in the mid-nineteenth century and is based on the premise that in logic only two possibilities exist: a statement is either true or false. Boolean algebra considers a true state to be 'on' and denoted by the symbol 1, the false state is 'off' and denoted by the symbol 0.

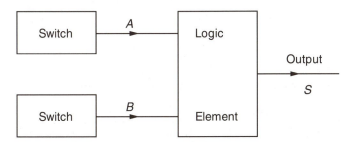

FIGURE 6.1 Electrical switches arranged to supply an AND gate.

Consider a switch. It may be either operated or not operated. When it is operated there is an output *A*, when it is not operated *A* is not there and can be denoted by NOT *A*. This is shown in Table 6.1.

TABLE 6.1 Logic conditions of a simple switch.

Switch condition		Output
Operated	1	*A*
Not operated	0	NOT *A*

NOT *A* is the inversion of *A* and is represented in Boolean algebra by showing a bar or line over the function which is inverted. So

$$\text{NOT } A = \bar{A}$$

and

$$\text{NOT } 1 = \bar{1} = 0$$

Basic logic elements

These are the AND and OR functions, plus their inversions NOT AND, known as NAND, and NOT OR, know as NOR.

Consider two switches *A* and *B* being fed into the logic element which has a logic output *S* (see Fig. 6.1).

Let the logic element be such that there is only an output when both inputs are present. Table 6.2 can be drawn showing the various combinations of inputs and outputs. This is known as a TRUTH TABLE. There is only an output when *A* and *B* are present, so an equation can be built up:

TABLE 6.2 Truth table for input *A* AND *B* and their respective output conditions.

A	*B*	*S*
0	0	0
0	1	0
1	0	0
1	1	1

$$A \text{ and } B \text{ gives } S$$

or, in Boolean algebra,

$$A . B = S$$

where the symbol . means AND. The inverse of S is NOT S and is obtained by inverting all the values of S.

NOT S is shown as $\bar{S}$, the bar denoting the inverse.

Expanding the previous truth table for the AND function to include $\bar{S}$ is shown in Table 6.3:

TABLE 6.3 Expanded truth table to include $\bar{S}$.

A	B	S	$\bar{S}$
0	0	0	1
0	1	0	1
1	0	0	1
1	1	1	0

$$S = A . B$$

Inverting

$$\bar{S} = \overline{A . B}$$

If the logic function used in Fig. 6.1 is such that an output is given when either A or B is present, then the truth table will be as shown in Table 6.4, the inverse is also shown:

TABLE 6.4 Truth table for input A OR B and their respective output conditions.

A	B	S	$\bar{S}$
0	0	0	1
0	1	1	0
1	0	1	0
1	1	1	0

$$S = A + B$$

where the logic function + denotes OR. Also

$$\bar{S} = \overline{A + B}$$

6.1.1 Rules for logic equation manipulation

As with ordinary algebra, certain rules can be applied to Boolean algebra. Consider the truth table, Table 6.5. There is an output S when A and $\bar{B}$ are present or when A and B are present. This may be represented algebraically as:

TABLE 6.5 Truth table for A and B conditions and their respective output conditions.

A	B	S	$\bar{S}$
0	0	0	1
0	1	0	1
1	0	1	0
1	1	1	0

$$S = A \cdot \bar{B} + A \cdot B$$

Brackets may be used as in normal algebra, thus

$$S = A \cdot (B + \bar{B})$$

A logic function can only have one of two states, either 1 or 0; if $B = 1$ then $\bar{B} = 0$ and if $B = 0$ then $\bar{B} = 1$. So $(B + \bar{B}) = 1$ whatever the state of B. Thus

$$S = A$$

Similarly,

$$\bar{S} = \bar{A} \cdot \bar{B} + \bar{A} \cdot B$$
$$= \bar{A}(B + \bar{B}) = \bar{A}$$

A set of rules can be developed by drawing up truth tables (Table 6.6) and comparing all the values of different functions.

TABLE 6.6 Truth table for all conditions of A and B.

A	B	$A + B$	$\overline{A + B}$	$A \cdot B$	$\overline{A \cdot B}$	$\bar{A}$	$\bar{B}$	$\bar{A} + \bar{B}$	$\overline{\bar{A} + \bar{B}}$	$\bar{A} \cdot \bar{B}$	$\overline{\bar{A} \cdot \bar{B}}$
0	0	0	1	0	1	1	1	1	0	1	0
0	1	1	0	0	1	1	0	1	0	0	1
1	0	1	0	0	1	0	1	1	0	0	1
1	1	1	0	1	0	0	0	0	1	0	1

Comparing the values in the columns it can be seen that

$$\overline{A + B} = \bar{A} \cdot \bar{B} \tag{1}$$

and

$$\overline{A \cdot B} = \bar{A} + \bar{B} \tag{2}$$

Inverting both sides of equation (1) gives

$$\overline{\overline{A + B}} = \overline{\bar{A} \cdot \bar{B}}$$

A double inversion restores it to the original state, hence

$$\overline{\overline{A + B}} = A + B = \overline{\bar{A} \cdot \bar{B}}$$

This can be seen in columns 5 and 10 in Table 6.6. Also

$$A \cdot B = \overline{\overline{A} + \overline{B}}$$

Summary of the rules of Boolean algebra

TABLE 6.7 Summary of rules of Boolean algebra.

	Theorem or rule
$A \cdot (A + B)$	$= A \cdot A + A \cdot B = A$
$A + A \cdot B$	$= A$
$A \cdot (\overline{A} + B)$	$= A \cdot B$
$A + \overline{A} \cdot B$	$= (A + B)$
$(A + B) \cdot (A + B)$	$= (A + B)$
$(A + B)(C + D)$	$= A \cdot C + A \cdot D + B \cdot C + B \cdot D$
$\overline{A} \cdot \overline{B}$	$= \overline{A + B}$
$A \cdot B$	$= \overline{\overline{A} + \overline{B}}$
$\overline{A} + \overline{B}$	$= \overline{A \cdot B}$
$A + B$	$= \overline{\overline{A} \cdot \overline{B}}$

The final four equations in Table 6.7 are known as De Morgan's Theorems and are useful to change AND functions to OR functions and vice versa.

6.2 Pneumatic valves as logic valves

Standard pneumatic valves will function as logic gates, giving an output when certain input conditions are met. Consider the 3-port valve shown in Fig. 6.2. There will only be an output S when both A and B are present.

$$S = A \cdot B$$

The valve is therefore acting as an AND gate.

A two-input AND gate may be illustrated using the symbols shown in Fig. 6.3.

A three-input AND gate using the same conventions may be shown using the symbols in Fig. 6.4.

A 3-port pneumatic valve can be used also as an OR gate by connecting it as shown in Fig. 6.5; its logic symbol is shown alongside.

There will be an output when either A or B is present. Thus

$$S = A + B$$

Specialist functions can be obtained by connecting the 3-port valve in different ways, as shown in Fig. 6.6.

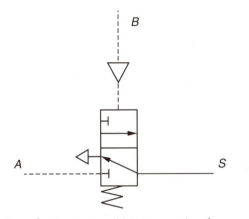

FIGURE 6.2 Application of a 3-port, 2-position pneumatic valve as an AND gate.

FIGURE 6.3 Symbolic representation of a two-input AND gate.

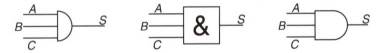

FIGURE 6.4 Symbolic representation of a three-input AND gate.

6.2.1 Comparison of logic and pneumatic symbols

The logic symbols –●– and ○– denote negation, and a + was also used to represent negation. So a NOR gate would have been as shown in Fig. 6.7.

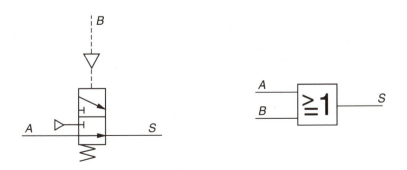

FIGURE 6.5 Application of a 3-port, 2-position pneumatic valve as an OR gate.

TABLE 6.8 Comparison of logic and pneumatic symbols.

Description and representation	Truth table	DIN 40700		US/Canadian	ISO 1219.1 pneumatic
		Old	New		
Amplification $S = X$	X S / 0 0 / 1 1				
Inversion (negation) $S = \bar{X}$	X S / 0 1 / 1 0				
OR $S = X + Y$ $S = X \vee Y$	X Y S / 0 0 0 / 0 1 1 / 1 1 1 / 1 0 1				
AND $S = X . Y$ $S = X \wedge Y$	X Y S / 0 0 0 / 0 1 0 / 1 1 1 / 1 0 0				
NOR OR gate with negated output $\bar{S} = X + Y$ $S = \overline{X+Y}$ $\bar{S} = X \vee Y$	X Y S / 0 0 1 / 0 1 0 / 1 1 0 / 1 0 0				
NAND AND gate with negated output $\bar{S} = X . Y$ $S = \overline{X . Y}$ $\bar{S} = X \wedge Y$	X Y S / 0 0 1 / 0 1 1 / 1 1 0 / 1 0 1				

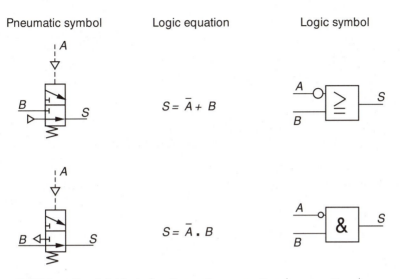

Pneumatic symbol	Logic equation	Logic symbol

$S = \bar{A} + B$

$S = \bar{A} \cdot B$

FIGURE 6.6 Specialist logic functions using conventional pneumatic valves.

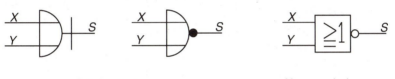

Old symbols   New symbol

FIGURE 6.7 Example of logic NOR symbols.

Where logic equations, i.e. Boolean algebra, occur in the script the signs + will be used for the OR function and . for the AND function. Elsewhere ∨ may be found for an OR function with ∧ being used for an AND function, the OR function may be confused with the letter Vee.

6.3 **Active and passive gates**

A passive device is one in which the output power is derived from the input signal. The output power will always be lower than the input power due to flow losses in the device.

Where a number of passive devices are used in series the output power (pressure) from succeeding devices will gradually reduce, until the output from one device is insufficient to operate the next. In a situation like this an amplifier must be used to boost the signal.

An active device is one in which power is continuously supplied and the output is derived from the input power and not the input signal.

The AND gate shown in Fig. 6.2 is a passive unit, but the OR gate shown in Fig. 6.3 is part active (when an input is at port B) and part passive ($\bar{B} \cdot A$).

FIGURE 6.8 NOR gate constructed using pneumatic valves.

Inverse functions

The NOR function is the inverse of the OR function, so there is no output when either *A* or *B* is present.

$$S = \overline{A + B}, \quad \text{i.e.} \quad \bar{S} = A + B$$

This can be achieved by using two 3/2 pneumatic valves connected as shown in Fig. 6.8, the logic symbol is shown alongside.

Similarly, the NAND gate gives no output when both *A* and *B* are present, as shown in Fig. 6.9.

Memory gates

A double-pilot-operated bistable valve can function as a memory. One pilot line is used to set or enable the valve, the other pilot resets or disables the valve. Truth tables are shown in Fig. 6.10 for both 3/2 and 5/2 double-pilot valves.

When the symbol ϕ is used it indicates an indeterminate state which will depend upon the last signal received. The signals can be either pulsed or continuous.

In the case of a 5/2 valve the memory function can be represented symbolically, as shown in Fig. 6.11.

The 5/2 valve may also be connected with two inputs; it will then function as a memory with two different outputs. This is shown in Fig. 6.12 together with its associated truth table.

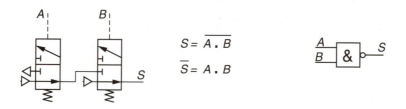

FIGURE 6.9 NAND gate constructed using pneumatic valves.

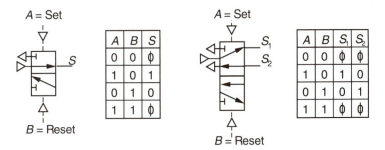

A	B	S
0	0	φ
1	0	1
0	1	0
1	1	φ

A	B	S_1	S_2
0	0	φ	φ
1	0	1	0
0	1	0	1
1	1	φ	φ

FIGURE 6.10 Pneumatic valves used as memory valves.

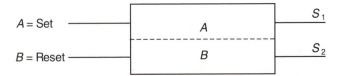

FIGURE 6.11 Symbolic representation of a memory function.

One application of this 5/2 valve is the group change valve in a cascade system, see Chapter 5.

6.4 Pneumatic logic valves

Specially designed pilot-operated miniature poppet valves giving a range of logic functions are available from several manufacturers. The 'Polylog' series manufactured by Compair

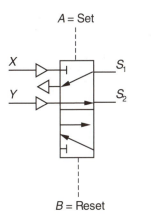

A	B	S_1	S_2
0	0	φ	φ
1	0	0	1
0	1	1	0
1	1	φ	φ

FIGURE 6.12 Pneumatic memory having two inputs and two outputs.

Function	Cross section	Symbol		Operation

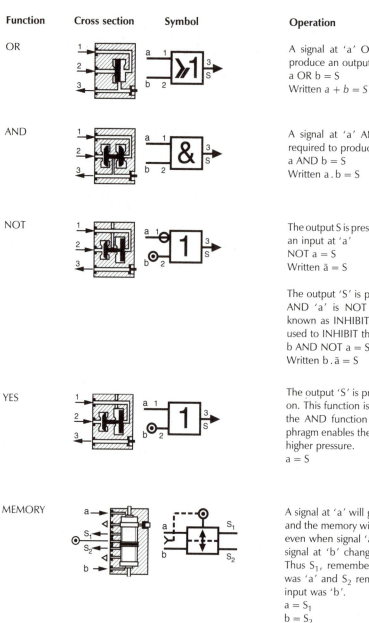

OR — A signal at 'a' OR a signal at 'b' will produce an output at 'S'
a OR b = S
Written $a + b = S$

AND — A signal at 'a' AND a signal at 'b' is required to produce an output at 'S'
a AND b = S
Written $a \cdot b = S$

NOT — The output S is present when there is NOT an input at 'a'
NOT a = S
Written $\bar{a} = S$

The output 'S' is present when 'b' is on AND 'a' is NOT on. This function is known as INHIBITION, i.e. signal 'a' is used to INHIBIT the path of signal 'b'
b AND NOT a = S
Written $b \cdot \bar{a} = S$

YES — The output 'S' is present when YES 'a' is on. This function is basically the same as the AND function except that the diaphragm enables the signal 'a' to switch a higher pressure.
a = S

MEMORY — A signal at 'a' will give an output at S_1 and the memory will remain in this state, even when signal 'a' is removed, until a signal at 'b' changes the output to S_2. Thus S_1, remembers that the last input was 'a' and S_2 remembers that the last input was 'b'.
$a = S_1$
$b = S_2$

FIGURE 6.13 'Polylog' logic components by Compair Maxam.

Maxam, shown in Fig. 6.13, consists of five basic units from which all the logic functions can be obtained. The 'Polylog' system has been designed to minimise pipework, the connection between logic elements being made in the valve base, with the only external connections being the input and output devices.

6.5 **Truth tables and their use**

Truth tables can be used to derive logic equations to solve various control problems. Once an equation has been established it can be reduced to its simplest form and translated into a circuit using logic valves.

Example 6.1

A door is to be opened by a pneumatic cylinder when either of two push-button valves A or B is operated. When the valve is released the door will close.

Solution

Let the inputs be A and B and let the output be S, so the door must open when $S = 1$.

Construct a truth table (see Table 6.9) listing all possible combinations of the inputs, and write down the required value of S for each of the inputs.

TABLE 6.9 Truth table for door opened by valves A and/or B and closed upon release of valves.

A	B	S	$\bar{S}$
0	0	0	1
1	0	1	0
0	1	1	0
1	1	1	0

Obtain from the truth table all the possibilities for $S = 1$.
S is equal to 1 when

$$A = 1 \quad \text{and} \quad B = 0$$

$$A = 0 \quad \text{and} \quad B = 1$$

$$A = 1 \quad \text{and} \quad B = 1$$

Using Boolean algebra an equation can be constructed:

$$S = A \,.\, \bar{B} + \bar{A} \,.\, B + A \,.\, B$$

where . means AND
  + means OR
  $A = 1$
  $\bar{A} = 0$

The equation may be simplified using normal algebraic methods to

$$S = A \cdot (\bar{B} + B) + \bar{A} \cdot B$$

The function of $A \cdot (\bar{B} + B)$ is dependent on A only, it does not matter if B is present or not. Therefore,

$$A \cdot (\bar{B} + B) = A$$

Thus,

$$S = A + \bar{A} \cdot B$$

which will further simplify to

$$S = A + B$$

from the rules of Boolean algebra (Table 6.7).

Alternatively, it is sometimes simpler to consider the value of $\bar{S}$, i.e. $S = 0$, and invert the value to give S, (i.e. $\bar{\bar{S}} = S$).

From the truth table

$$\bar{S} = \bar{A} \cdot \bar{B}$$

From Table 6.7

$$\bar{A} \cdot \bar{B} = \overline{A + B} \quad \text{(De Morgan)}$$

Thus

$$\bar{S} = \overline{A + B}$$

By inverting both sides, $S = A + B$ as before.

A pneumatic circuit obtained from this equation would contain one OR gate. A possible solution using a double-acting cylinder and a 5/2 control valve is shown in Fig. 6.14.

Expand the example so that the doors are to be opened by changing the state of A and B and closed by changing the state of A and B. This is the same problem as two light switches controlling one light on the stairway in a house. The first stage in the solution is to draw up a truth table as before. Assuming that with both A and B in the off condition, S is off; if the opposite starting assumption were made the solution would be identical.

TABLE 6.10 Truth table for opening and closing the door by changing the state of A and B.

A	B	S	$\bar{S}$
0	0	0	1
1	0	1	0
1	1	0	1
0	1	1	0

Thus the door is closed if both A and B are operated or unoperated.

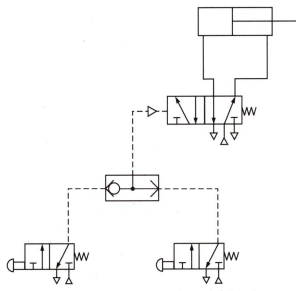

FIGURE 6.14 Pneumatic circuit incorporating a shuttle valve as an OR gate.

Extract the equation for S:

$$S = A \cdot \bar{B} + \bar{A} \cdot B$$

This equation will not simplify and the logic circuit using AND, OR gates is shown in Fig. 6.15.

Alternatively, write down the equation for $\bar{S}$:

$$\bar{S} = \bar{A} \cdot \bar{B} + A \cdot B$$

Invert

$$S = \overline{\bar{A} \cdot \bar{B} + A \cdot B}$$

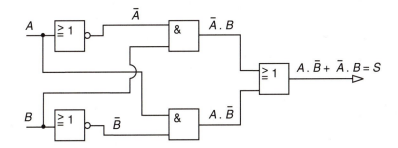

FIGURE 6.15 Logic symbol diagram for the equation $(A \cdot \bar{B}) + (\bar{A} \cdot B) = S$.

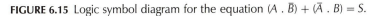

From De Morgan

$$\bar{A} \cdot \bar{B} = \overline{A + B}$$

So

$$S = \overline{A + B} + A \cdot B$$

This equation can be translated into a logic circuit, as shown in Fig. 6.16.

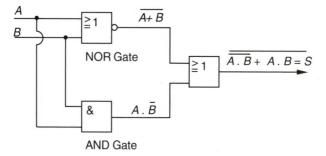

FIGURE 6.16 Logic symbol diagram for the equation $\overline{A + B} + A \cdot B = S$.

It should be noted that by working with the inverse function of S the solution has been reduced from five to three logic gates. This particular function is known as exclusive OR and some manufacturers produce integrated logic blocks to give this function.

6.5.1 Complement of functions

It is sometimes possible to simplify logic functions by using the 'complement' of the function. The complement of a single variable is the opposite: the complement of 1 is 0; the complement of A is $\bar{A}$.

The complement of a Boolean function is obtained by

changing all   . to +

changing all   + to .

changing all   1 to 0

changing all   0 to 1

and complementing all laterals.

Consider

$$S = A \cdot B \cdot \bar{C} + C \cdot 1 + 0 \cdot \bar{A} \cdot \bar{B}$$

Complement the function

$$S_{\text{comp}} = \bar{S} = (\bar{A} + \bar{B} + C) \cdot (\bar{C} + 0) \cdot (1 + A + B)$$

Consider

$$S = A + B$$
$$S_{\text{comp}} = \bar{S} = \bar{A} \cdot \bar{B}$$

Invert

$$\bar{\bar{S}} = \overline{\bar{A} \cdot \bar{B}}$$

So

$$S = \overline{\bar{A} \cdot \bar{B}} = A + B$$

Similarly, $S = A \cdot B$

$$S_{\text{comp}} \bar{S} = \bar{A} + \bar{B}$$

Invert

$$\bar{\bar{S}} = S = \overline{\bar{A} + \bar{B}} = A + B$$

These statements are De Morgan's Theorem as shown in Table 6.7.

6.6 Karnaugh maps

This is a method of representing the logic variables in a diagram which consists of a series of squares, with each square allocated to one particular combination of variables. The chart is arranged so that adjacent squares, both horizontally and vertically, differ only by one variable. 1, 2 and 3 variable maps are shown in Fig. 6.17.

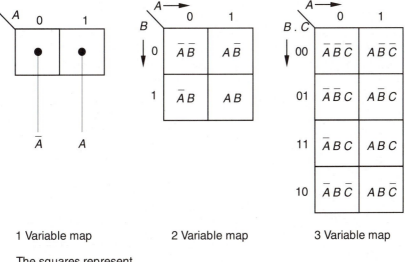

1 Variable map   2 Variable map   3 Variable map

The squares represent
$\bar{A}$ and A

FIGURE 6.17 One-, two- and three-variable Karnaugh maps.

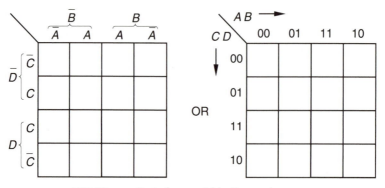

FIGURE 6.18 Basic four-variable Karnaugh map.

The variable represented by each square is shown written in that square in Fig. 6.17; this is only shown as an illustration and is not normally done. It should be noted that the Karnaugh map can be considered to be drawn on a sphere so that the top and bottom end lines coincide, as do the extreme left- and right-hand lines. It is very important to note how the squares are designated so that there is only a change of one variable in each adjacent square. A four-variable map showing all the possible combinations of the logic variables *A*, *B*, *C* and *D* is shown in Fig. 6.18 as a 4 × 4 array, giving 16 squares.

The greater the number of variables the greater the number of squares required on the Karnaugh map.

$$\text{Number of squares} = 2^n$$

where *n* is the number of variables

So, for a five variable function, 2^5 squares are needed or an 8 × 4 map. If the three variables are shown horizontally it will require eight columns which must only alter by one variable in adjacent boxes. It may be easier to consider how each variable changes, as shown in Fig. 6.19.

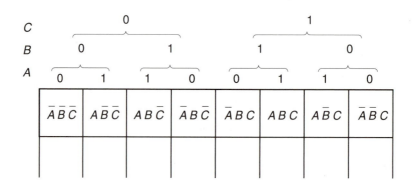

FIGURE 6.19 Diagram to show how each variable changes.

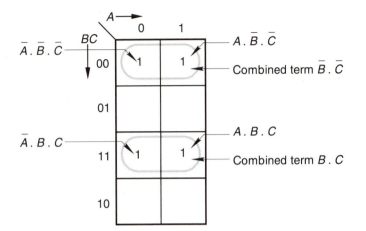

FIGURE 6.20 Karnaugh map representing the equation $A \cdot B \cdot C + \bar{A} \cdot B \cdot C + \bar{B} \cdot \bar{C}$.

6.6.1 Simplification using Karnaugh maps

If a function has two terms in adjacent squares it means that only one variable has changed values in those two squares and that that variable can be eliminated from the function. Consider the function

$$S = A \cdot B \cdot C + \bar{A} \cdot B \cdot C + \bar{B} \cdot \bar{C}$$

The term $A \cdot B \cdot C$ is represented by one square on the Karnaugh map, as is the term $\bar{A} \cdot B \cdot C$, whereas the term $\bar{B} \cdot \bar{C}$ covers two squares as the term is independent of A, i.e.

$$\bar{B} \cdot \bar{C} = \bar{B} \cdot \bar{C} \cdot (A + \bar{A}) = A \cdot \bar{B} \cdot \bar{C} + \bar{A} \cdot \bar{B} \cdot \bar{C}$$

The Karnaugh map for the function will be as shown in Fig. 6.20. Therefore, the function

$$S = A \cdot B \cdot C + \bar{A} \cdot B \cdot C + \bar{B} \cdot \bar{C}$$

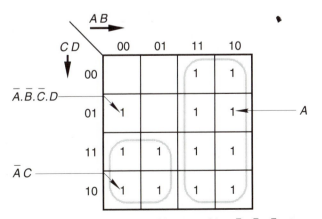

FIGURE 6.21 Karnaugh map of four variables. $\bar{A} \cdot \bar{B} \cdot \bar{C} \cdot D$.

Simplifies to

$$S = B \cdot C + \bar{B} \cdot \bar{C}$$

Consider the four variable Kanaugh map in Fig. 6.21.

The item $\bar{A} \cdot \bar{B} \cdot \bar{C} \cdot D$ contains all four variables and will only occupy one square on the map; however, the term $\bar{A} \cdot C$ only contains two variables and covers four squares, whereas the term A contains one variable and covers eight squares.

6.6.2 Karnaugh map for minimisation

In Karnaugh maps adjacent squares vary by one term only. The minimisation technique is to locate and loop adjacent squares as pairs, blocks of four, eight, etc., which are then read as a single term consisting only of variables having the same value in all the looped squares.

Example 6.2

Minimise the function:

$$S = A \cdot \bar{B} + \bar{A} \cdot \bar{B} \cdot \bar{C} + A \cdot B \cdot C \cdot D$$
$$+ \bar{A} \cdot \bar{B} \cdot \bar{C} \cdot D$$

Methods

1. Show the terms on a Karnaugh map.
2. Loop adjacent terms into blocks of two, four, eight, etc.
3. Write down simplified expressions which consist of all the variables having the same value in the block, (Fig. 6.22).

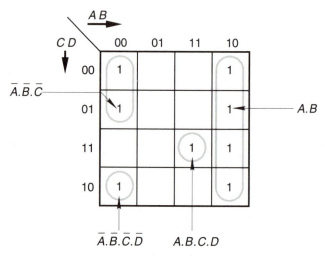

FIGURE 6.22 Karnaugh map for minimisation.

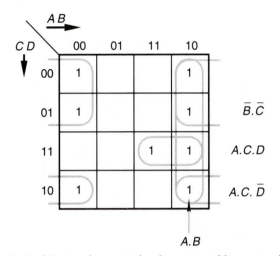

FIGURE 6.23 Revised Karnaugh map with adjacent variables grouped together.

The map has been redrawn (Fig. 6.23) with adjacent variables grouped into blocks of two and four.

Note: The left- and right-hand boundaries of the map are the same, as are the upper and lower boundaries. Variables can be used as many times as needed to simplify the expression. So,

$$S = A . B + \bar{B} . \bar{C} + A . C . D + A . C . \bar{D}$$

Example 6.3

A machined plate shown in Fig. 6.24 has two holes A and B drilled in it. A pneumatic inspection machine is used to check the presence of the holes and the length of the plate.

Pneumatic sensors W, X, Y, Z are used to detect the holes and check the length of the plate. The sensor signals are

$$W = \text{hole at } A, \quad \bar{W} = \text{no hole at } A$$

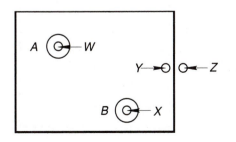

FIGURE 6.24 Machined plate.

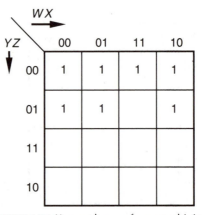

FIGURE 6.25 Karnaugh map for remachining.

$$X = \text{hole at } B, \bar{X} = \text{no hole at } B$$

$$Y = \text{plate too short}$$

$$\bar{Z} = \text{plate too long.}$$

For the plate to pass the inspection the following conditions must be fulfilled:

$$\text{Correct} = W \cdot X \cdot \bar{Y} \cdot Z$$

The plate can be reworked, i.e. corrected, if a hole is missing or if the plate is too long.

$$\text{Remachine} = \bar{W} \cdot X \cdot \bar{Y} \cdot Z + \bar{W} \cdot \bar{X} \cdot \bar{Y} \cdot Z$$

$$+ \bar{W} \cdot X \cdot \bar{Y} \cdot \bar{Z} + \bar{W} \cdot \bar{X} \cdot \bar{Y} \cdot \bar{Z}$$

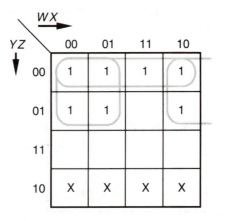

FIGURE 6.26 Karnaugh map for remachining including impossible outputs.

$$+ W . X . \bar{Y} . \bar{Z} + W . \bar{X} . \bar{Y} . \bar{Z}$$

$$+ W . \bar{X} . \bar{Y} . Z$$

If the plate is short it has to be scrapped.

$$\text{Scrap} = W . X . Y . Z + \bar{W} . X . Y . Z + \bar{W} . \bar{X} . Y . Z$$

$$+ W . \bar{X} . Y . Z$$

It is impossible for the sensor outputs to be equal to $W . X . Y . \bar{Z} + \bar{W} . X . Y . \bar{Z} + W . \bar{X} . \bar{Y} . \bar{Z} + \bar{W} . \bar{X} . Y . \bar{Z}$, as this indicates that the plate is both too short and too long, an impossibility.

To simplify the expressions, draw a Kanaugh map for the remachine condition (Fig. 6.25).

The outputs which are impossible can be used to minimise the expression for remachine. Draw the Kanaugh map for remachine and include the impossible outputs (Fig. 6.26).

Group the signals into twos, fours or eights. Three groups of four will include all the remachine outputs without using the impossible combinations. Thus

$$\text{Remachine} = Y . Z + W . Y + X . Y$$

Consider the scrap condition and draw the Karnaugh map including the impossible combinations of outputs (Fig. 6.27).

$$1 = \text{scrap}$$

$$X = \text{impossible}$$

Minimising

$$\text{Scrap} = Y$$

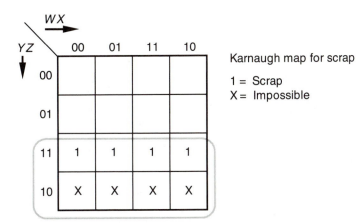

FIGURE 6.27 Karnaugh map including scrap and impossible output conditions.

Now

$$\text{Correct} = W \cdot X \cdot \bar{Y} \cdot Z$$

$$\text{Scrap} = Y$$

All other possibilities must be capable of being remachined. Thus

$$\text{Remachine} = \overline{W \cdot X \cdot \bar{Y} \cdot Z + Y}$$

Depending upon the logic gates available, the expression for remachine may well be simpler than the first expression obtained.

6.7 Sequential control

A sequential control system is a system in which the outputs or operations are required to change in a predetermined order.

Sequential systems can be divided into two groups:

1. *Synchronous*, in which the next output state depends only on the present output state and an input pulse. A counter circuit is a typical example. Synchronous circuits may be considered as time-based systems.
2. *Asynchronous*, in which one operation can only begin when the previous operation has been completed. An asynchronous circuit is an event-based system.

A typical asynchronous system could be a cylinder sequence, for example:

$$A + \ B + \ A - \ B -$$

represented by the circuit shown in Fig. 6.28.

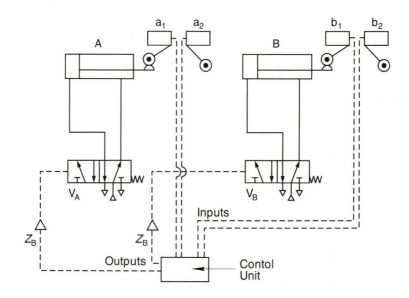

FIGURE 6.28 Typical asynchronous cylinder circuit.

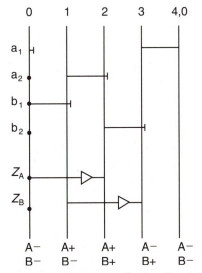

FIGURE 6.29 Step timing chart for $A+$ $B+$ $A-$ $B-$.

The operation of the sequence can be shown in several ways and the logic for the control unit derived from the various methods.

Timing chart

The sequence is divided into steps, which indicate changes in the variables but not equal intervals of time (Fig. 6.29). The output Z_A must be present while the cylinder A is moving forward or remains extended, i.e. between lines 0 and 2. Therefore

$$Z_A = b_1 + a_2 . b_2$$

Similarly,

$$Z_B = a_2 + b_2 . a_1$$

Mealy diagrams are used to show the way in which a circuit changes state and can be useful in formulating a problem. Consider the sequence used previously, i.e. $A+$ $B+$ $A-$ $B-$.

The circles in Fig. 6.30 represent the various states of the circuit; the changes in state are shown by the arrowed lines. The letters on the lines represent the input combinations causing the change and the output combination resulting. A more usual way of using the Mealy diagram is shown for the same sequence (Fig. 6.31) but using binary code to represent the condition of the trip valves and the output signals.

Karnaugh maps are a very useful method of showing the sequence of a circuit. Consider the previous sequence: points a_1 and a_2 represent cylinder A retracted and extended. Similarly, b_1 and b_2 represent the condition of cylinder B.

These points can be represented on a grid (Fig. 6.32(a)). A square on the grid represents the end state of both cylinder A and cylinder B, as shown in Fig. 6.32(b).

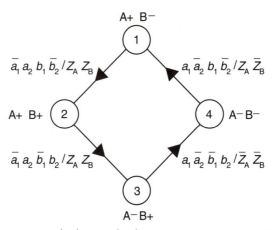

FIGURE 6.30 Mealy diagram for the sequence $A+ B+ A- B-$.

Note that adjacent squares horizontally or vertically only differ by the movement of one cylinder. As an alternative to using a_1 and a_2, etc., to indicate the cylinder state, 0, 1 are sometimes used to show the condition of a_1 and a_2, so the Karnaugh map becomes that shown in Fig. 6.33.

Consider the sequence $A+ B+ B- A-$ on a Karnaugh map (Fig. 6.34). The start condition for the sequence can be indicated by the letter S.

If the cylinders are operated by bistable devices, e.g. double-pilot-operated valves, a pulse only is required to change the valve from one state to the other. The logic signal to cause A to extend is $A+ = S . b_1$.

Check the squares on the Karnaugh map covered by b_1 to see if any other square is in the $A-$ condition, b_1 covers squares 1 and 2, but $A-$ only exists in square 1. So $A+ = S . b_1$ is unique. Similarly $B+ = a_2$.

Check that a_2 occurs in squares 2 and 4 and $B-$ in squares 1 and 2; thus $B+ = a_2$ is unique. Similarly, $A- = b_2$ and $B- = a_1$.

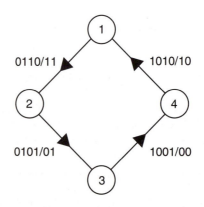

FIGURE 6.31 Mealy diagram using binary code for trip valves.

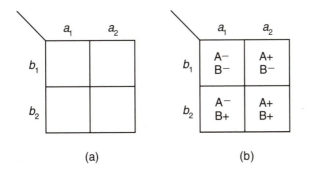

FIGURE 6.32 (a) Basic grid form; (b) grid showing end of state for both cylinders.

Consider a larger network using three cylinders with the sequence

$$A+ \ B+ \ C+ \ A- \ C- \ B-$$

Draw a Karnaugh map for three terms (Fig. 6.35) showing the start condition on the first square, i.e. a_1, b_1, c_1.

Depending upon the way in which the diagram is drawn, a different plot results. Consider the network shown in the diagram at the top. To extract the logic equations, consider the previous operation, the completion of which initiates the next stage. Ensure that each logic statement is unique and if necessary, introduce an additional term.

Consider the A+ signal which follows B−:

$$A+ = b_1$$

Inspect the squares under b_1, i.e. 1, 2, 5 and 6, and see if any are in the A− condition. Square 1 is the only square in b_1 and the A− condition, a start signal S must also be included.

Therefore

$$A+ = b_1 . S$$

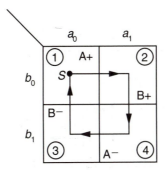

FIGURE 6.33 Modified Karnaugh map.

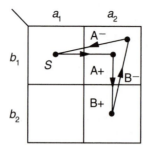

FIGURE 6.34 Karnaugh map for the sequence A+ B+ B− A−.

The next operation B+ is signalled by the completion of A+, i.e., B+ $= a_2$. Now a_2 covers squares 2, 3, 6 and 7, but only square 2 is B−. Therefore

$$B+ = a_2$$

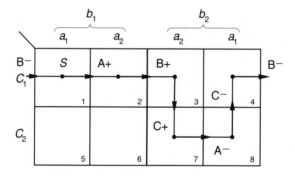

OR

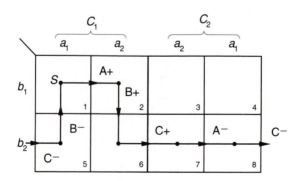

FIGURE 6.35 Karnaugh map for the sequence A+ B+ C+ A− B− C−.

Next is C+, signalled by b_2, i.e.

$$C+ = b_2$$

But b_2 covers squares 3, 4, 7 and 8 and C is in minus conditions in squares 3 and 4, so the expression $C+ = b_2$ is not unique and another term must be added. The difference between 3 and 4 is the condition of cylinder A. Therefore

$$C+ = b_2 . a_2$$

The next term is A−, signalled by C_2, so $A- = c_2$ and c_2 covers squares 5, 6, 7 and 8 but the only square in use in which A+ (a_2) and c_2 exist is number 7. Therefore

$$A- = c_2$$

Next is C−, signalled by a_1. Now a_1 and c_2 (C+) cover squares 5 and 8, but only square 8 is in use. Therefore

$$C- = a_1.$$

Finally, B− is signalled by c_1; squares 3 and 4 are in use and contain c_1 and b_2 (B+). The difference between these squares is the condition of A. Therefore

$$B- = c_1 . a_1$$

This completes the logic equation and a circuit can now be drawn.

6.7.1 Secondary functions

If, in a particular sequence when a Karnaugh map is drawn, one square is used twice, a memory or second function must be used to differentiate between the two routes through the one square. With wall attachment elements or miniature spool valves, flip flops or bistable devices are the most economical memories. When poppet valves or spring units are used at least two elements are needed to form the memory.

Consider the sequence A+ B+ B− A−. The Karnaugh map is shown in Fig. 6.36.

A secondary function Y must be introduced after B+ to prevent the circuit going into oscillation. The sequence becomes

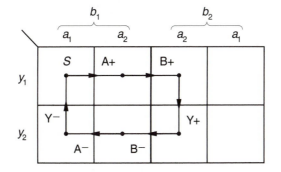

FIGURE 6.36 Karnaugh map for the sequence A+ B+ B− A− incorporating a secondary function Y.

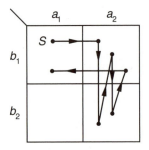

FIGURE 6.37 Karnaugh map for the sequence $A + B + B - B + B - A -$.

$$A+ \quad B+ \quad Y+ \quad B- \quad A- \quad Y-$$

The secondary function map is shown in Fig. 6.36.
The logic equations can now be found.

$$A+ = y_1 \cdot S \quad \text{Unique}$$

$$B+ = a_2 \quad \text{Not unique introduce condition of Y}$$

Therefore,

$$B+ = a_2 \cdot y_1 \quad \text{Unique}$$

$$Y+ = b_2 \quad \text{Unique}$$

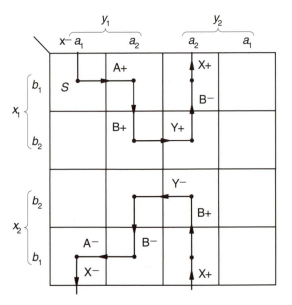

FIGURE 6.38 Secondary function Karnaugh map for the sequence $A+ \quad B+ \quad Y + B - X + B + Y - B - A - X-$.

$$B- = y_2 \qquad \text{Unique}$$

$$A- = b_1 \qquad \text{Not unique introduce condition of Y}$$

$$A- = b_1 \cdot y_2 \quad \text{Unique}$$

$$Y- = a_1 \qquad \text{Unique}$$

Repeating operations

Consider the sequence

$$A+ \ B+ \ B- \ B+ \ B- \ A-$$

The Karnaugh map is shown in Fig. 6.37.

The $a_2 b_1$ and $a_2 b_2$ squares are used more than once, which means that secondary signals must be incorporated. The sequence then becomes

$$A+ \ B+ \ Y+ \ B- \ X+ \ B+ \ Y- \ B- \ A- \ X-$$

The secondary function map for this sequence is shown in Fig. 6.38.

If a double output memory device is not available a single output unit having output states $Y+$ and $Y-$ in place of y_1 and y_2 can be used.

To extract the logic equations, proceed as previous examples:

$$A+ = x_1 \cdot S \qquad \text{Unique}$$

$$B+ = a_2 \qquad \text{Not unique}$$

Therefore

$$B+ = a_2 \cdot x_1 \quad \text{Unique}$$

$$Y+ = b_2 \qquad \text{Not unique}$$

$$Y+ = b_2 \cdot x_1 \quad \text{Unique}$$

$$B- = y_2 \qquad \text{Not unique}$$

$$B- = y_2 \cdot x_2 \quad \text{Unique}$$

$$X+ = b_1 \qquad \text{Not unique}$$

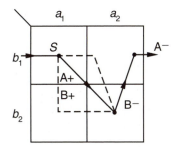

FIGURE 6.39 Karnaugh map for two operations (A+ B+) at the same time.

$$X+ = b_1 . y_2 \quad \text{Unique}$$

$$B+ = x_2 \quad \text{Not unique}$$

$$B+ = x_2 . y_2 \quad \text{Unique}$$

$$Y- = b_2 \quad \text{Not unique}$$

$$Y- = b_2 . x_2 \quad \text{Unique}$$

$$B- = y_1 \quad \text{Not unique}$$

$$B- = y_1 . x_2 \quad \text{Unique}$$

$$A- = b_1 \quad \text{Not unique}$$

$$A- = b_1 . y_1 \quad \text{Unique}$$

$$X- = a_1 \quad \text{Unique}$$

The circuit diagram can now be drawn for this sequence.

6.7.2 Multiple operations

If two or more operations begin at the same time the path line on the Karnaugh map can be diagonal (Fig. 6.39).

Example 6.4

Sequence

$$A+ \quad B- \quad A-$$
$$B+$$

In this case the map would be as shown in Fig. 6.40.

The operations $A+$ and $B+$ starting simultaneously may result in $A+$ being completed first, in which case the path will be square 1 to square 2 to square 4, but if $B+$ is completed first the path will be square 1 to 3 then to 4.

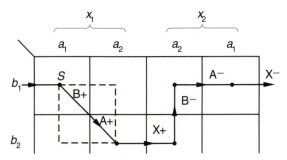

FIGURE 6.40 Karnaugh map for the sequence $A + B - A - B +$.

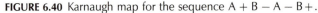

Although squares 2 or 3 may be occupied *en-route* there are many occasions when these can be taken as empty squares and the next operation, B$-$, can proceed directly into square 2. In this case the logic will be:

$$A+ \text{ and } B+ = S \cdot a_1, x_1$$

$$X+ \qquad = a_2 \cdot b_2 \cdot x_1$$

$$B- \qquad = x_2$$

$$A- \qquad = a_2 \cdot b_1, x_2$$

$$X- \qquad = a_1 \cdot x_2$$

6.7.3 Multiple sequences and multiple actions

Some circuits are required to operate in more than one sequence depending upon the input signals; for example, an automatic drilling unit, the normal cycle being:

$$A+ \ B+ \ C+ \ C- \ B- \ \text{Repeat}$$
$$A-$$

where   A is the feed cylinder
B is the clamp cylinder
C is the drill cylinder.

A sensor *d* indicates if a component is correctly positioned. If so, the sequence continues; if not, the machine returns to the start condition and stops, sounding an alarm. Thus the sequence is as shown in Fig. 6.41. The cylinder A layout is shown in Fig. 6.42.

The step involving the movements C$+$ and A$-$ at the same time can be considered as a diagonal movement on the Karnaugh map, provided a false signal cannot result due to one trip operating before the other. If this is possible a secondary sequence has to be used as previously described.

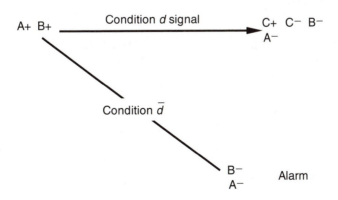

FIGURE 6.41 Alternative routes sequences.

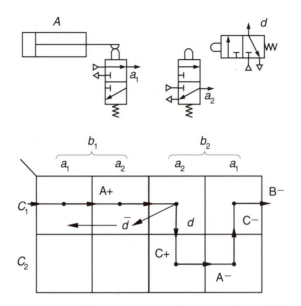

FIGURE 6.42 Diagrammatic layout of an automatic drill unit together with the Karnaugh map.

In this particular circuit there are two possible routes of returning to the start condition and a secondary circuit must be used to differentiate between them, the condition of *d* controlling the secondary circuit (Fig. 6.43).

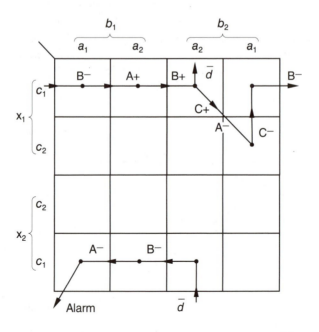

FIGURE 6.43 Karnaugh map of secondary circuit to overcome false signals.

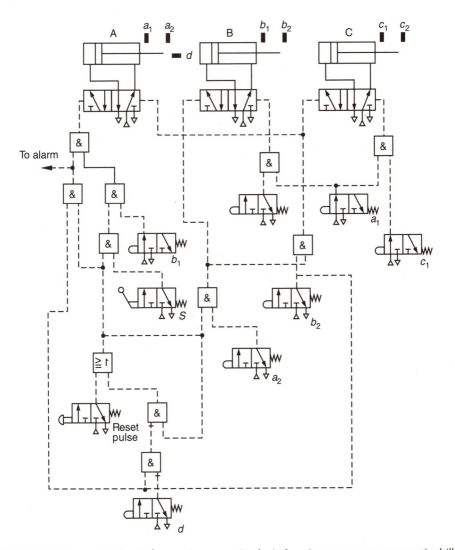

FIGURE 6.44 Pneumatic circuit diagram incorporating logic functions to operate automatic drill.

The logic becomes:

$$A+ = S \cdot b_1 \cdot x_1$$

$$B+ = a_2 \cdot x_1$$

$$C+ \text{ and } A- = b_2 \cdot a_2 \cdot x_1$$

$$C- = c_2 \cdot a_1$$

$$B- = c_1 \cdot a_1$$

Alternatively,

$$A+ = S \cdot b_{1_{,}} \cdot x_1$$

$$B+ = a_2 \cdot x$$

$$X+ = b_2 \cdot \bar{d}$$

$$A- \text{ and } B- = b_2 \cdot x_2$$

The complete circuit for this sequence is shown in Fig. 6.44.

CHAPTER 7

Electro-pneumatics

The electrical control of pneumatic circuits, or electro-pneumatics, offers some advantages over the use of pilot air signalling. The response speed of the control circuit is very much faster as the electrical signal travels almost instantaneously through the circuit. By eliminating the air pilots and the pipe runs, the air consumption of the circuit is reduced. There may well be an increase in the cost of the system as solenoid-operated valves are more expensive than pneumatically piloted valves; this, however, is offset to some extent by the saving in pipework. The use of low-voltage control circuits, either alternating current (a.c.) or direct current (d.c.), improves the safety aspects.

Electrical components used in pneumatic circuits include switches, solenoids, relays, timers, sensors and transducers.

7.1 Switches

Almost every piece of electrical equipment includes a switch of some type. Switches are available in a wide range of sizes, designs and operations. The factors that influence the selection of a switch are listed below.

Current and voltage rating

1. Current- or power-switching rating is that current or power that may be switched in a purely resistive circuit.
2. Voltage rating is the maximum voltage that can be carried by the switch in a purely resistive circuit.
3. Current-carrying rating is the maximum current that can flow through the closed contacts. This is normally higher than the current-switching rating.

Contact configuration

1. Make, break or changeover action.
2. If the changeover action required is 'make before break' or 'break before make'.
3. The number of poles in the switch (this is the number of separate circuits that can be switched simultaneously).
4. The number of ways or throws (the number of positions to which each pole may be switched).
5. Type of switch action – momentary action which resets the switch when the operating force is removed.

6. Latching action in which the switch remains in the operated position even when the operating force is removed; to reset the switch a reset force must be applied.

Examples

1. Single pole, single throw, single-pole on/off. Abbreviated to SPST.
2. Single pole, double throw, single-pole changeover. Abbreviated to SPDT
3. Double pole, single throw, double-pole on/off. Abbreviated to DPST.
4. Double pole, double throw, double-pole changeover. Abbreviated to DPDT.

Method of switch operation

Switch operation may be by push button, lever or rotate. The action of a switch involves making or breaking an electrical current, the action of the switch when breaking direct and alternating currents is different. The electrical arc created or drawn when contacts carrying a d.c. current are parted will vaporise the contact metal and maintain an arc over a considerable distance. Thus for d.c. switching a rapid break and wide contact separation is required. In an alternating current the voltage falls to zero twice per cycle, and the arc drawn on switching a.c. tends to be extinguished.

An inductive load may cause current and voltage surges as the circuit is broken and the magnetic flux collapses. With capacitive load there is a high initial current surge as the circuit is made. With both inductive and capacitive load the switches should be de-rated by about 20 per cent otherwise the switch life can be considerably reduced.

Switches which are suitable for a.c. and d.c. systems are given two ratings; for example, 250 V a.c.6A or 30 V d.c.10A.

7.1.1 Limit switches

In pneumatic systems, as in many others, it is essential to know that one operation is complete before the next starts. Switches can be used for this purpose, indicating that a aguard is closed, a component is in place, a cylinder has fully extended and so on. There are two basic types of switch

1. Mechanical contacting.
2. Proximity or non-contacting.

Mechanical limit switches

These are operated by a moving part striking the actuating mechanism of the limit switch. There are many types of actuators (plunger, roller, lever, etc.). The majority of limit switches have a pair of normally open contacts and a pair of normally closed contacts, but different numbers of contacts and different configurations are available.

When selecting a limit switch for a particular application the following points must be considered: Contact rating, a.c. or d.c.; current life expectancy of the contacts; contact bounce and how it can affect any control circuit; physical size of the switch; movement required to operate switch; effect of over-run; effect of environment on switch; and the protection needed.

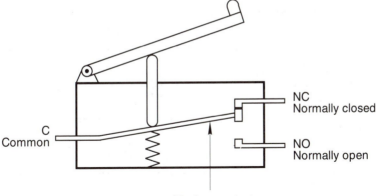

FIGURE 7.1 Basic construction of a micro-switch.

Micro-switches are a particular type of limit switch. They are smaller in size and have only three contacts, a common contact (C), a normally open contact (NO) and a normally closed contact (NC), as in Fig. 7.1. When the switch is operated the contacts change over, the NO contact is made and the NC contact is broken.

Non-contacting switches

A number of different types are available, the most common being

- magnetic
- inductive
- capacitive
- photo-electric.

Magnetic or reed switches consist of a pair of reeds with silver- or gold-plated contacts sealed in a glass envelope which is filled with an inert gas. When a magnet or magnetic field is brought close to the reeds it causes the reeds to become magnetised and either attract or repel each other, so changing the contacts over. Proximity switches are available, built into pneumatic cylinders; reed switches are positioned on the cylinder barrel and the piston is fitted with magnets.

Inductive proximity switches are solid-state devices, i.e. no moving parts. They are switched by the proximity of a metal object. The switching element is either a transistor or a thyristor and can handle a limited electrical power. If the unit is overloaded it will rapidly short out. It is essential that correct selection is made.

Capacitive proximity switches are similar to the inductive type but are capable of sensing both metallic and non-metallic objects. Again careful selection of a switch for the particular duty is essential and it may well be advisable to consult the manufacturer.

Two types of output are available from proximity switches: NPN or PNP. In the NPN type the load is connected between the positive supply and the proximity switch output, whereas in the PNP type the load is connected between the negative supply and the output. This is shown in Fig. 7.2.

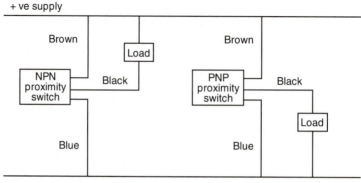
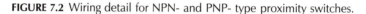

−ve supply

FIGURE 7.2 Wiring detail for NPN- and PNP- type proximity switches.

Proximity switches are particularly suitable for interfacing with programmable logic controllers (PLCs) and may be connected directly to input terminals.

Optical sensors utilise a light source or emitter and a receiver. The light source may be infrared, white light or polarised light; it is essential that ambient light cannot cause a malfunction. In normal operation the transmitter and receiver can be up to 5 metres apart depending on the type of unit. Where very long distances are involved, laser light can be employed. When selecting an optical sensor the environment must be considered carefully, as a very dusty atmosphere can soon clog the transmitter with a coat of dirt. When space is at a premium, fibre-optics can be utilised. As the switching device is solid state all the precautions mentioned in the proximity switches apply.

7.2 **Solenoids**

In pneumatics there are two major types of solenoid:

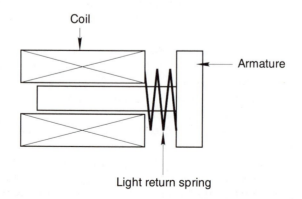

FIGURE 7.3 Typical construction of a solenoid.

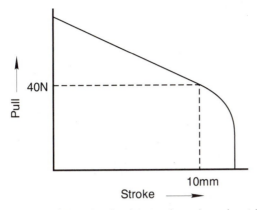

FIGURE 7.4 Force/stroke diagram for a d.c. solenoid.

1. a.c. solenoids
2. d.c. solenoids

The principle of operation of both types of solenoid is similar. An electrical coil wound on a former, laminated for a.c. coils, surrounds an armature, as shown in Fig. 7.3.

When the coil is energised, an electro-magnetic field forms which pulls the laminated steel armature into the coil. The pull exerted on the armature depends upon the electrical power applied to the coil and the stroke of the armature. A typical force–stroke curve for a solenoid is shown in Fig. 7.4.

The force reduces rapidly as the voltage reduces. The shorter the stroke the greater the pull available. The solenoid characteristics in Fig. 7.4 show that a force of 40 N is available at a stroke of 10 mm. When the armature is at the end of the stroke the pull will be considerably greater than 40 N; this will cause 'hammering' unless some form of cushioning is used.

The maximum permissible frequency of operation of a solenoid depends upon the stroke and force required; the less the force or stroke the higher the maximum operating frequency. Figure 7.5 shows the relationship between current flowing through a coil and stroke, assuming a constant load.

The current flowing in the d.c. coil is almost constant, independent of stroke. However, the initial current in the a.c. coil is very high. This is known as the in-rush current, and is up to 10 times the current flowing when the armature is at the end of its travel, i.e. at zero stroke. An a.c. coil must always be allowed to complete its stroke, otherwise the large current flowing will burn out the coil.

Solenoids have two main applications in electro-pneumatic systems:

1. To operate valves, i.e. solenoid valves
2. To operate electrical relays.

7.3 **Electrical relays**

These are electrically operated switches which may be single or multiple. A diagrammatic sketch of a relay is shown in Fig. 7.6.

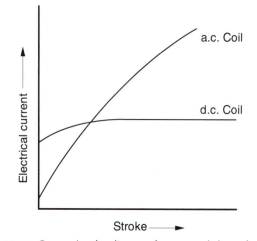

FIGURE 7.5 Current/stroke diagram for a.c. and d.c. solenoids.

Relays are available with either a.c. or d.c. coils to operate at voltages up to about 240 V d.c., 440 V a.c. The operating voltage of the coil is not influenced by the voltage of the relay contact switch. The solenoid coil may be 12 V d.c. and be switching 120 V a.c. or more.

The life expectancy of a relay depends upon the switch contact material and is usually quoted at a given current and voltage for a.c. or d.c. Relays are available with a fixed number of poles, so many normally open and so many normally closed; with some types the number of poles can be altered by adding contact blocks as needed. The relay contacts can be arranged to be normally open, normally closed, changeover, or make before break.

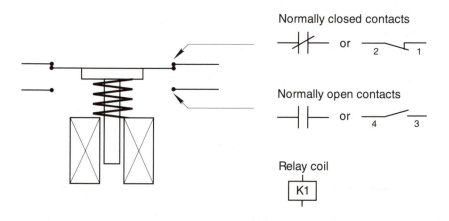

FIGURE 7.6 Relay construction.

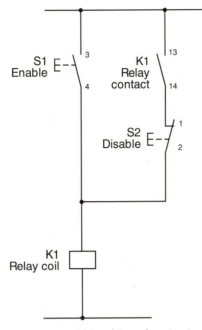

FIGURE 7.7 (a) Latching relay circuit.

7.3.1 Latching or hold-on relay circuits

If a relay is operated by a pulse it is necessary for the relay to remain energised even when the pulse is removed, so a latching circuit is used. Two basic latching circuits are shown in Figs. 7.7(a) and 7.7(b).

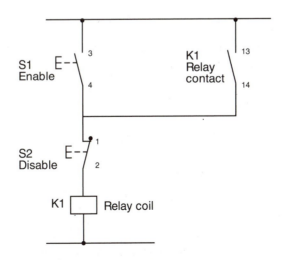

FIGURE 7.7 (b) Latching relay circuit.

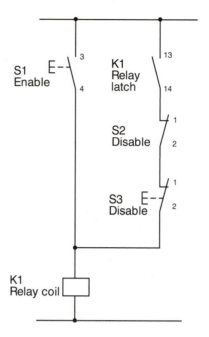

FIGURE 7.8 Start/stop circuit.

Switch S1 energises the relay coil K1 which changes the relay contacts over; relay contact K1.1 closes and keeps the relay coil energised even when the switch S1 is opened. K1.1 is the latching contact.

The reset signal S2 is used to switch the relay off. The reset signal has to be logically inverted to give the disabled signal or, in logic terms,

$$\text{Relay disabled} = K_d = \bar{K} = \overline{S2}$$

The term $\overline{S2}$ means the normally closed contacts of S2. When these contacts are changed the circuit is broken and the relay is de-energised.

Example 7.1

If the 'Relay on' signal is S1 and the 'Relay off' signal is S2 or S3, draw the electrical circuit for the relay (see Fig. 7.8).

$$\text{Relay enable} \quad K1e = S1$$

$$\text{Relay disable} \quad K1d = \overline{S2 + S3}$$

$$= \overline{S2} \, . \, \overline{S3}$$

Note: S2 and S3 are normally closed contacts. The use of Boolean algebra is explained in Chapter 6.

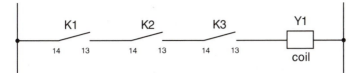

FIGURE 7.9 Open relay contacts in series as a logic AND circuit.

7.3.2 Time delay relays

A circuit using a resistor and capacitor will give time delays after the input is applied before the output voltage reaches a certain value. The output voltage can be made to trigger off a circuit to energise a relay coil by using a transistor. These timing devices can be very accurate with very good repeatability.

7.3.3 Relay logic

Relays have a number of normally open and normally closed contacts which can be interconnected to form AND and OR logic functions (see Fig. 7.9).

The coil will be energised when relay contacts K1, K2 and K3 are all operated. This is a three-input AND gate, that is,

$$K1 . K2 . K3 = Y1$$

If a normally closed contact is used for, say, K2 the circuit is as shown in Fig. 7.10 and the logic equation becomes

$$K1 . \overline{K2} . K3 = Y1$$

Similarly, if the contacts are arranged in parallel, the circuit will be as shown in Fig. 7.11. The coil will be energised if K1 or K2 or K3 are operated. The logic equation is

$$K1 + K2 + K3 = Y1$$

By using a combination of series and parallel connections, normally open and normally closed contacts, all logic functions can be obtained. See Chapter 6 for further details.

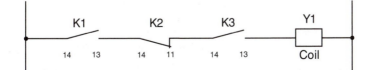

FIGURE 7.10 Open and closed relay contacts in series as a logic AND circuit.

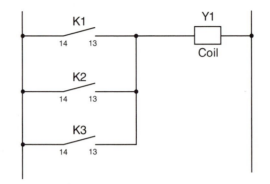

FIGURE 7.11 Open relay contacts in parallel as a logic OR circuit.

7.4 **Solenoid valves**

In the earlier designs the solenoid acted directly on the valve spool moving it from one position to another, the spool being returned either by a spring or a second solenoid. The solenoid had to exert a considerable thrust over a relatively long stroke, which necessitated a high-power solenoid. When a.c. solenoids are used a very heavy initial current flows to start the solenoid moving and if, for any reason, the solenoid cannot complete its stroke, the heavy current will damage the coil. To overcome this problem it is common to use the solenoid to operate a small air pilot valve, the pilot air then being applied to move the main spool. A typical solenoid-operated pilot valve is shown in Fig. 7.12.

The pilot valve is arranged to operate the main spool valve, as shown symbolically in Fig. 7.13.

The air supply to the pilot stage may be internally connected from the main valve or it may need a separate supply as shown.

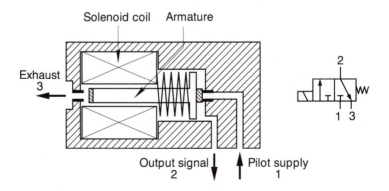

FIGURE 7.12 Solenoid operated 3-port, 2-position pilot valve.

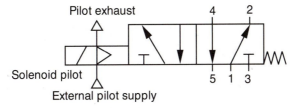

FIGURE 7.13 Symbolic representation of a solenoid-controlled pilot-operated direction control valve.

Single-solenoid valves

These are usually monostable devices which are returned to their original condition by a spring when the solenoid is de-energised. With this type of valve due regard must be taken of the possibility of failure of the control circuits, which would reset all the single solenoid valves and could cause the actuators to operate in a dangerous manner. Valve reset on failure of the electrical control circuit can be designed into the system as a failsafe feature.

The response times of valves are defined as:

1. *Response time: energise*
   The time from the switching command 'On' to when the pressure at the output port of the valve rises to approximately 20 per cent of the nominal pressure at a supply pressure of 6 bar and a temperature of 20 °C.
2. *Response time: de-energise*
   This is the time from the switching command 'Off' to the pressure at the output port falling to 10% of the nominal pressure (starting to fall in the case of a 2-port, 2-position valve) at a supply pressure of 6 bar and a temperature of 20 °C.

Single solenoid air reset valves are bistable devices which remain in one state until either an electrical signal or air pilot changes that state.

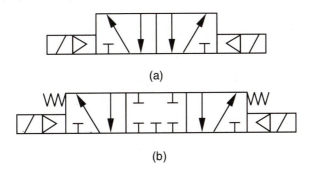

FIGURE 7.14 Symbolic representation of (a) a bistable 5/2 valve; (b) a monostable 5/3 direction control valve.

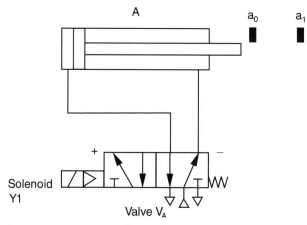

FIGURE 7.15 (a) Basic electrically controlled cylinder circuit.

Double-solenoid valves

A 2-position double-solenoid valve, as shown in Fig. 7.14(a), is a bistable device that will remain in a given state until it receives a signal to change its state. A 3-position spring centred valve (Fig. 7.14(b)) will always return to the centre position when the solenoids are de-energised, and is therefore a monostable valve.

7.5 Electrical control circuits

7.5.1 Control of single-solenoid valves

Consider the pneumatic system in Fig. 7.15(a) with limit switches a_0 and a_1. The limit switches are shown in their state at the start of the sequence. For example, limit switch a_0 is shown in the operated condition, indicating that cylinder A is retracted.

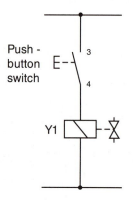

FIGURE 7.15 (b) Electrical circuit to give cylinder extend and retract operation under manual control.

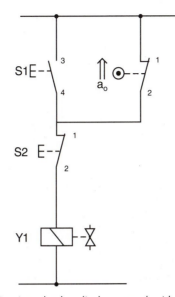

FIGURE 7.15 (c) Electrical circuit to latch cylinder extend with manual de-latching switch for retract.

The circuit can be made to function in many different modes depending upon the electrical control circuit. To maintain the cylinder extending, or in the extend condition, the solenoid must remain energised.

The control circuit in Fig. 7.15(b) will cause the cylinder to extend when the push-button is operated, and retract as soon as it is released. The cylinder will not necessarily fully extend.

In Fig. 7.15(c) switch S1 is normally open and switch S2 is normally closed. The retract limit switch a_0 is shown in its operated condition when the cylinder is fully retracted. Operating switch S1 will energise the solenoid A causing the cylinder to extend, de-energising the limit switch a_0. Releasing S1 will not de-energise solenoid A as the normally closed contact on a_0 will be made, connecting the supply to solenoid A. The cylinder will continue extending or remain extended until switch S2 is operated, breaking the circuit to solenoid A. The cylinder will retract as long as switch S2 is operated; if the switch is released before the cylinder retracts to operate the limit switch a_0, the solenoid is re-energised and the cylinder will extend.

7.5.2 Relay circuits

Automatic operation can be achieved using relay circuits, as shown in Fig. 7.16.

Relay K1 is energised when the extend (enable) switch S1 is operated. Relay contact K1.1 (which is normally open) is closed when the relay is energised, and this contact maintains a supply to relay K1 even if S1 is released. The second relay contact, K1.2, also closes, energising solenoid Y1, which in turn will cause the cylinder to extend. When the extend limit switch a_1 is operated, the normally closed contacts open, de-energising the relay. This causes the solenoid Y1 to be de-energised and the cylinder retracts. If

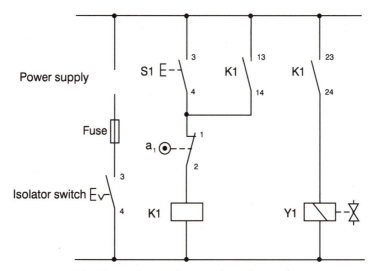

FIGURE 7.16 Automatic operation using a relay.

switch S1 remains operated the cylinder will oscillate about the extend position. This problem can be overcome by using limit switch a_0 in series with switch S1, so that S1 will only make the circuit to the relay when limit switch a_0 is operated. This circuit is shown in Fig. 7.17.

In Figs 7.16 and 7.17 two sets of relay contacts are used; one set to latch the relay and the second to energise the solenoid. Provided the current-carrying capacity of the

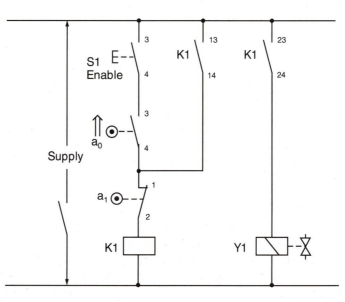

FIGURE 7.17 Electrical circuit to prevent oscillation in the extend position if the start switch remains operated. Note: ⇑ indicates that the switch is operated.

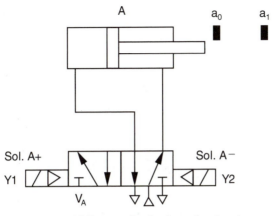

FIGURE 7.18 (a) Pneumatic circuit to give A+ A−.

contacts is sufficient, one set could be used to both latch the relay and operate the solenoid.

The cylinder can be made to reciprocate continuously by using the circuit shown in Fig. 7.17 and keeping S1 operated or by substituting an on/off bistable switch for S1. When the switch is operated the cylinder reciprocates, on releasing or de-energising switch S1 the cylinder will complete the cycle and stop in the fully retracted position.

7.5.3 Control of double-solenoid 2-position valves

The pneumatic circuit for a single cylinder controlled by a double-solenoid valve is shown in Fig. 7.18(a). Again different modes of operation can be obtained, dependent on the type of control circuit. The valve shown in Fig. 7.18(a) is a solenoid-controlled pilot-operated valve. When the solenoid is energised an air signal is applied to pilot the main spool over; should both solenoids be energised a pilot signal is applied to both sides of the spool and the spool will move to or remain in the position to which the first

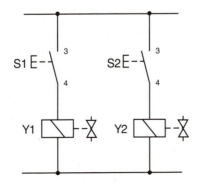

FIGURE 7.18 (b) Manual control to give A+ A−.

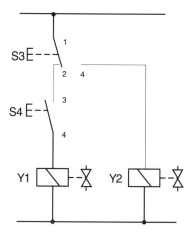

FIGURE 7.18 (c) Interlocking electrical circuit to give A+ A−.

pilot signal switched it. The solenoids will not be damaged. In the case of directly operated a.c. solenoid valves in which the solenoid force acts directly on the main spool, both solenoids must not be energised together otherwise the coils will overheat and burn out. This does not occur in d.c. solenoids as previously discussed.

In the electrical circuit in Fig. 7.18(b), operating switch S1 energises solenoid A+ and switch S2 energises solenoid A−; operating S1 and S2 will energise both solenoids, which may not be desirable. Figure 7.18(c) shows how this can be avoided by using a simple electrical interlock.

In the circuit shown in Fig. 7.18(c), operating switch S3 will energise solenoid A− independent of the state of S4, whereas to energise solenoid A+ switch S3 must be unoperated and switch S3 operated. This can be expressed in logic terms as

$$\text{Sol. A}- = \text{S3}$$

$$\text{Sol. A}+ = \overline{\text{S3}} . \text{S4}$$

The solenoids are only energised to switch the valve over, the pilot pressure being applied to the spool to switch the valves. This is sometimes referred to as 'pressure applied' operation and is the method most frequently used.

Pressure release operation, as the name implies, maintains pilot air pressure on both sides of the spool, the release of the pilot pressure on one side of the spool allows the remaining pilot to switch the valve over. The electrical circuit in Fig. 7.19 will give this effect, both solenoids remaining energised until the push-button is operated. The electrical circuit is similar to Fig. 7.18(b) except that normally closed contacts are used on the switches.

An interlock circuit can be made similar to Fig. 7.18(c) to ensure that both solenoids cannot be de-energised at the same time.

Automatic return and continuous reciprocation of the cylinder can be achieved by using the limit switches a_0 and a_1 in Fig. 7.18(a) to replace the push-button switches shown in Figs 7.18(b) and 7.18(c). A stop/run switch must also be included to enable

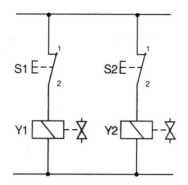

FIGURE 7.19 Electrical circuit for the control of a pressure release operation.

the cylinder to be stopped in either the retracted or extended position, depending upon the positioning of the stop/run switch in the circuit.

The electrical circuit for continuous reciprocation is shown in Fig. 7.20. At the side of the operating rungs in the ladder circuit, the operation occurring, e.g. cylinder A+, has been shown, although in this case it is obvious. In more complex sequences this type of notation can be very useful.

7.6 Multi-cylinder circuits

Consider a two-cylinder circuit using the same nomenclature as previously, the valve controlling cylinder A being valve V_A. When solenoid Y2 is energised cylinder A is retracted and when solenoid Y1 is energised cylinder A is extended. Similarly, when

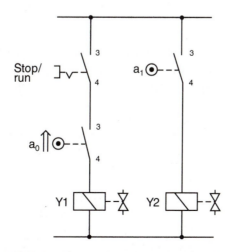

FIGURE 7.20 Automatic reciprocating circuit.

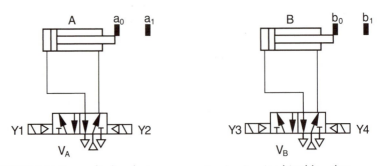

FIGURE 7.21 Two-cylinder electro-pneumatic circuit using bistable valves.

solenoid Y4 is energised cylinder B is retracted and when solenoid Y3 is energised cylinder B is extended. The pneumatic circuit with the appropriate limit switches is shown in Fig. 7.21.

Again the electrical control circuit will determine the operating sequence of the cylinders. Consider the cylinder sequence $A+$, $B+$, $A-$, $B-$ repeat. The $A+$ operation is to occur when cylinder A is fully retracted, cylinder B is fully retracted and the stop/run switch is in the run condition. The $A-$ condition is signalled by limit switch a_0 and $B-$ by b_0. Thus

$$A+ \text{ signal} = \text{Stop/run switch, } a_0 \text{ and } b_0$$

However, in this sequence the $B-$ signal is generated by the $A-$ operation, so b_0 will not be operated unless a_0 is also operated; thus, the simplified expression becomes

$$A+ \text{ signal} = \text{Stop/run switch and } b_0$$

or, as a logic equation

$$A+ = S1 \cdot b_0$$

where S1 is the stop/run switch.

Table 7.1 can be constructed giving details of each step in the sequence; for a simple sequence, as used, it is not essential, but will be useful for more complex sequences. The electrical ladder diagram shown in Fig. 7.22 can be drawn from the table.

TABLE 7.1 Operating sequence and events $A+$ $B+$ $A-$ $B-$ repeat.

Operation	Solendoid	Initiated by	Completion signal
$A+$	Y1	S1 AND b_1 ($S1 \cdot b_1$)	a_1
$B+$	Y2	a_1	b_1
$A-$	Y3	b_1	a_0
$B-$	Y4	a_0	b_0

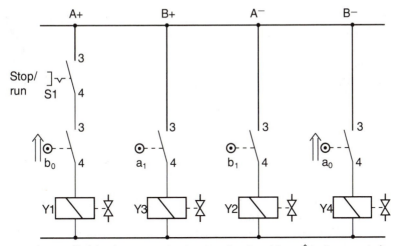

FIGURE 7.22 Electrical ladder diagram to give A+ B+ A− B−. Note: ⇑ indicates switch operated at end of sequence.

In Fig. 7.22 the rungs in the ladder start on the left and move to the right, representing the order of the sequence. This is not necessarily so in all ladder diagrams; the order of the rungs does not have to be in the order of the sequence and, indeed, it is quite common to group the operations of each cylinder together, as shown in Fig. 7.23, which gives an identical sequence to Fig. 7.22 and is identical electrically.

7.6.1 Cascade electrical circuits

Consider the sequence A+ A− B+ B−. If the same method were used, the A+ solenoid, which is energised by the B− limit switch, would remain energised when the A− solenoid was energised by the A− limit switch. There would be an electrical signal on both sides

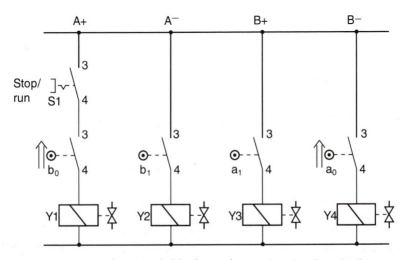

FIGURE 7.23 Alternative ladder layout for A+ B+ A− B− circuit.

of valves V_A, resulting in a trapped signal similar to that mentioned in pneumatic sequences. One of the methods of overcoming the problem was to use a 'cascade' circuit; a similar method can be used in the electrical circuit.

Divide the sequence into groups of letters so that no letter appears more than once in any group, with relays being used to select the appropriate group. The sequence can be written

	A+	A− B+	B−
Group	I	II	I
Relay	$\overline{K1}$	K1	$\overline{K1}$

Only one relay is used to give the two groups, the relay being de-energised for group I and energised for group II. Note that at the commencement of the sequence all the group change relays MUST be de-energised. Table 7.2 can be drawn giving the operation, method of initiation and so on, including the relay changes. The completion of one operation initiates the next (a group change is considered as an operation).

TABLE 7.2 Operating sequence and events A+ B+ B− A− including the cascade relay

Operation	Solenoid or relay	Group and relay	Initiated by	Completion signal
A+	Y1	I $\overline{K1}$	S1 . b_0 . $\overline{K1}$	a_1
A−	Y2	II K1	K1	a_0
B+	Y3	II K1	a_0 . K1	b_1
B−	Y4	I $\overline{K1}$	$\overline{K1}$	b_0
Group II K1 (enable)	K1		a_1	
Group I K1 (disable)	K1		b_1	

S1 is the stop/run switch
$\overline{K1}$ denotes NOT K1 or K1 disabled
K1 denotes K1 enabled

The relay K1 has to be enabled and disabled as it is a monostable device. The relay disable signal is the inverse of the $\overline{K1}$ signal. This is fully explained in the chapter on pneumatic logic.

To have an output signal from $\overline{K1}$, use the normally closed contacts numbered 1 and 2.

Consider the electrical ladder circuit diagram shown in Fig. 7.24, which has been obtained from considering Table 7.2. This circuit can be redrawn to reduce the number of relay contacts used and to simplify the wiring, as shown in Fig. 7.25.

The same technique can be used on multi-cylinder sequences or on repeat function sequences. Consider the two cylinder sequence

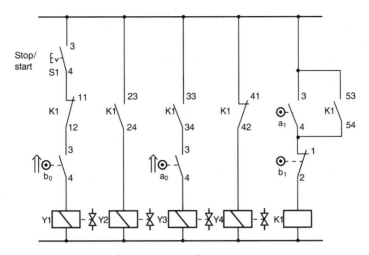

FIGURE 7.24 Ladder diagram to give the sequence A+ A− B+ B−.

A+ B+ B− A− B+ B− repeat

A sequence must be divided into two groups so that no letter repeats in any group.

	A+ B+	B− A−	B+	B−
Group	I	II	III	IV
Relays	$\overline{K1}\ \overline{K2}$	$K1\ \overline{K2}$	$K1\ K2$	$\overline{K1}\ K2$

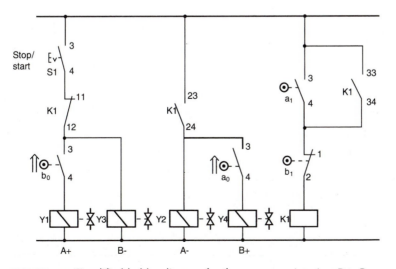

FIGURE 7.25 Simplified ladder diagram for the sequence A+ A− B+ B−.

To obtain the four groups only two relays, K1 and K2, have been used, the normally open and normally closed contacts of the relays being utilised to create four groups. In group I, both relays are in the de-energised condition, only the state of one relay being changed in successive groups. Using the cylinder layout as shown in Fig. 7.21, construct an operation table (Table 7.3).

TABLE 7.3 Operating sequence and events $A+\ B+\ B-\ A-\ B+\ B-$

Operation	Solenoid or relay	Group and relay	Initiated by	Completion signal
A+	Y1	I $\overline{K1} . \overline{K2}$	S1 . $\overline{K1} . \overline{K2}$	a_1
B+	Y2	I $\overline{K1} . \overline{K2}$	$a_1 . \overline{K1} . \overline{K2}$	b_1
Relay K1	K1	I $\overline{K1} . \overline{K2}$	$b_1 . \overline{K1} . \overline{K2}$	K1
B−	Y3	II K1 . $\overline{K2}$	K1 . $\overline{K2}$	b_0
A−	Y4	II K1 . $\overline{K2}$	$b_0 . K1 . \overline{K2}$	a_0
Relay K2	K2	II K1 . $\overline{K2}$	$a_0 . K1 . \overline{K2}$	K2
B+	Y2	III K1 . K2	K1 . K2	b_1
Relay K1	K1	III K1 . K2	$b_1 . K1 . K2$	$\overline{K1}$
B−	Y3	IV $\overline{K1}$. K2	$\overline{K1}$. K2	b_0
Relay K2	K2	IV $\overline{K1}$. K2	$b_0 . \overline{K1} . K2$	$\overline{K2}$

In the ladder circuit (Fig. 7.26) the solenoids Y3 and Y4 have been shown twice to simplify the drawing of the diagram. The actual circuit for solenoid Y3 would be as shown partially in Fig. 7.27. The circuit for solenoid Y4 is similar.

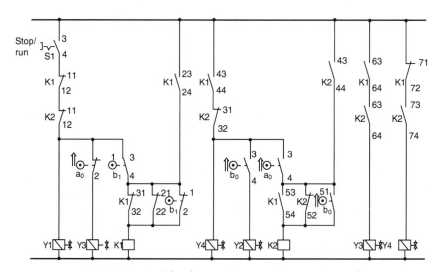

FIGURE 7.26 Ladder diagram for a repeat operation cycle.

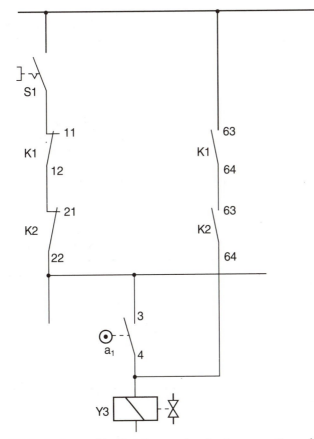

FIGURE 7.27 Partial ladder diagram showing two operations of solenoid Y3.

Limit switches b_0 and b_1 have also been used twice. If there is any possibility of operation of the relays causing a malfunction of the sequence, limit switches with two completely separate sets of contacts must be used.

7.6.2 Use of single-solenoid valves in multi-cylinder circuits

The preceding multi-cylinders have used double-solenoid 5/2 valves. The sequences can be obtained using single-solenoid spring return 5/2 valves which are held energised using relays. Consider the pneumatic circuit shown in Fig. 7.28. Should there be a failure of the electrical control supply, both cylinders will retract.

Draw the operational table for the sequence A+ B+ A− B− repeat (Table 7.4). There are no trapped signals in this sequence, so a cascade circuit does not have to be used but a relay is needed for each solenoid.

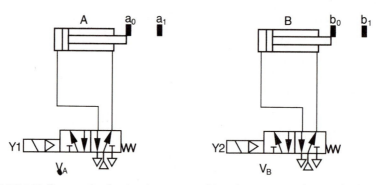

FIGURE 7.28 Pneumatic circuit using monostable valves to control two cylinders.

TABLE 7.4 Operating sequence and events $A+$ $B+$ $A-$ $B-$ using single-solenoid valves.

Operation	Solendoid or relay	Initiated by	Completion signal
$A+$	$K1 . Y1$	$S1 . b_0 . K1$	a_1
$B+$	$K2 . Y2$	$a_1 . K2$	b_1
$A-$	$\overline{K1} . \overline{Y1}$	$b_1 . \overline{K1}$	a_0
$B-$	$\overline{K2} . \overline{Y2}$	$a_0 . \overline{K2}$	b_0

Note:

$$\text{Relay K1 disable signal} = \overline{K1-}$$

$$= b_1$$

Similarly,

$$\text{Relay K2 disable signal} = \overline{K2-}$$

$$= a_0$$

The electrical circuit may now be drawn, as shown in Fig. 7.29.

7.6.3 Concurrent operation of steps in a sequence

In the previous sequence the two cylinders are required to start to retract at the same time in order to reduce the overall cycle time. The sequence can be written as

$$A+ \ B+ \ A-$$
$$B-$$

Draw the operational table for this sequence (Table 7.5).

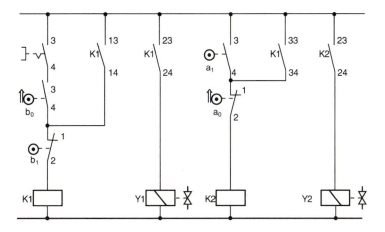

FIGURE 7.29 Ladder diagram for the control of two monostable valves to operate two cylinders in the sequence A+ B+ A- B−.

TABLE 7.5 Operating sequence and events A+ B+ A−.
B−

Operation	Solenoid or relay	Initiated by	Completion signal
A+	K1	$S1 . a_0 . b_0$	
	Y1	K1	a_1
B+	K2	a_1	
	Y2	K2	b_1
A−⎫	$\overline{K1} . \overline{K2}$	b_1	
B−⎭	$\overline{Y1} . \overline{Y2}$	$\overline{K1} . \overline{K2}$	b_0

The disable signal for both relay K1 and relay K2 is the inverse of the K− signal, i.e.

$$\text{K1 disable} = \text{K2 disable} = \overline{b_1}$$

The electrical circuit for the sequence is shown in Fig. 7.30. One disable contact $\overline{b_1}$ is used to disable both relays.

Relay K1 contacts 3 & 4 latch relay K1 and energise solenoid Y1. Relay K2 contacts 3 & 4 latch relay K2 and energise solenoid Y2.

7.6.4 Cascade circuits using single-solenoid valves

The pneumatic circuit shown in Fig. 7.20 is required to operate in the following sequence:

$$A+ B+ B- A- \text{ repeat}$$

To avoid trapped signals in this sequence a cascade circuit must be used. First divide the sequence into groups so no letter repeats in a group. In this case, there are two groups:

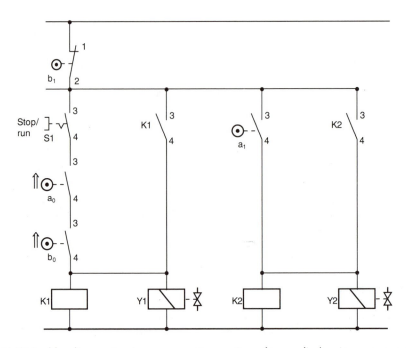

FIGURE 7.30 Ladder diagram to give concurrent operation of two cylinders in a sequence.

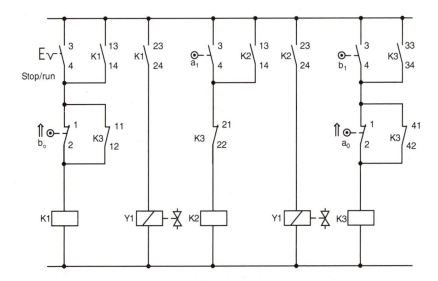

FIGURE 7.31 Ladder diagram to provide a cascade operation using single-solenoid valves.

$$A+, B+ \quad | \quad B-, A-$$

$$I \qquad \qquad II$$

$$\overline{K3} \qquad \quad \overline{K3}$$

Use a relay K3 to act as a group change relay. When the relay is de-energised group I is supplied with a control electrical supply, and when relay K3 is energised group II is supplied. Relays K1 and K2 are used to control solenoids Y1 and Y2 respectively. Draw up the operational scheme (Table 7.6).

TABLE 7.6 Operating sequence and events $A+ B+ B- A-$ using single-solenoid valves.

Operation	Solenoid or relay	Group and relay	Initiated by	Completion signal
$A+$	K1	I $\overline{K3}$	$S1 . \overline{K3}$	
	Y1		K1	a_1
$B+$	K2	I $\overline{K3}$	$a_1 . \overline{K3}$	
	Y2		K2	b_1
Group change relay K3	K3	I $\overline{K3}$	$b_1 . \overline{K3}$	K3
$B-$	$\overline{K2}$	II K3	K3	
	$\overline{Y3}$		$\overline{K2}$	b_0
$A-$	$\overline{K1}$	II K3	$b_0 . K3$	
	$\overline{Y4}$		$\overline{K1}$	a_0
Group change relay K3	$\overline{K3}$	II K3	$a_0 . K3$	$\overline{K3}$

The relay disable signals are as follows:

$$K1 \text{ disable} = \overline{K1-} = \overline{b_0 . K3} = \overline{b_0} + \overline{K3}$$

$$K2 \text{ disable} = \overline{K2-} = \overline{K3}$$

$$K3 \text{ disable} = \overline{K3-} = \overline{a_0 . K3} = \overline{a_0} + \overline{K3}.$$

The ladder circuit may now be drawn and is shown in Fig. 7.31.

Circuits for sequences with more than two cylinders and with repetitive operations can be constructed by using similar techniques.

CHAPTER 8

Programmable Logic Controllers

The programmable logic controller (PLC) is a solid-state device designed to perform the logic functions previously accomplished by electro-mechanical relays, mechanical/pneumatic timers and counters, drum switches, sequence controllers, etc. The PLC offers several advantages over electro-mechanical relay systems. The ever-increasing complexity of modern production systems demands faster, more flexible and more reliable control systems. Relay control systems have to be hard wired to perform a specific function; should system requirements change then the relay wiring has to be changed or modified. In extreme cases, complete control panels will have to be changed since rewiring may be uneconomical. Programmable controllers were first introduced into industry in the late 1960s. They have since become much more sophisticated, smaller and cheaper and are rapidly replacing the traditional hard-wired systems.

The block diagram shown in Fig. 8.1 represents the basic structure of a PLC. The core of the unit houses the central processing unit (CPU) or 'Brain' of the system in the form of a micro-processor. Besides the micro-processor, the processor contains memory and communication chips to enable it to interface with the programmer, printer and other PLCs.

The micro-processor carries out three functions:

1. To monitor the state of the input devices.
2. To solve the logic of the user program.
3. To set the output devices as required.

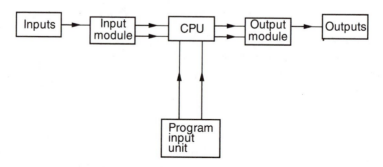

FIGURE 8.1 Block diagram of a basic PLC structure.

8.1 Programming devices

The programming device is usually a separate device used to enter a program into the PLC. This program will determine the sequence of operation of the system. These devices are generally in the form of a hand-held unit with either a light-emitting diode (LED) display or a liquid crystal display. Providing that the correct software is available, then a computer can be used to program the PLC.

8.1.1 Input modules

The input module accepts signals from sensors and/or switches and converts the signals into a form acceptable to the CPU. The input module has a number of input points, each with its own specific address. The input module circuits have optical isolators to separate input voltages from the logic circuits, thereby preventing damage to the processor if high voltages are inadvertently connected to the input module. Opto-isolators can also help to reduce the effects of electrical noise which can cause erratic processor operation.

The input module selects the required inputs and instructs the CPU to carry out certain operations on the inputs to give the required output signal. This signal is fed to the output module which has a series of output connections, each with its own address.

8.1.2 Output modules

The output module has three types of output device: volt free, transistorised and triacs.

Volt-free outputs

Volt-free outputs are relay contacts from internal relays within the PLC. These can switch a wide range of output voltages and currents making them suitable for inductive loads (e.g. solenoids) and resistive loads (e.g. indicator lamps). If the current rating of the output relay is insufficient a secondary relay must be used.

Transistorised outputs

Transistorised outputs use a solid-state device to switch the outputs. This limits the outputs to 24 V d.c. and maximum currents of 2 A per block (of 8 outputs). If these are to be used a separate external power supply is required.

Triacs

Triacs are a second solid-state device which can be used to switch solenoids and contactor coils directly. Triacs do not have ON or OFF states but have HIGH or LOW resistance states. When in the high resistance state a small leakage current will flow which causes no problem with solenoid coils.

8.2 Application of PLCs

PLCs can be used to control any system that can be operated by making or breaking a pair of contacts. The application of PLCs with pneumatic systems will now be considered, and may be represented diagrammatically in Fig. 8.2.

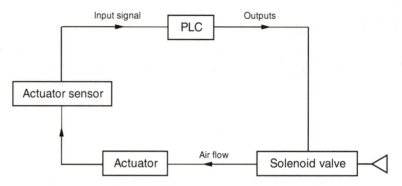

FIGURE 8.2 Diagrammatic representation of the application of a PLC.

The actuator can be a pneumatic cylinder, motor or semi-rotary actuator. The actuator may be energised until

(a) a certain position is reached, signalled by a position sensor;
(b) the air supply or the air discharging from the actuator is a certain pressure, signalled by a pressure switch;
(c) the actuators have completed a number of complete cycles, signalled by limit switches and an internal counter in the PLCs;
(d) the actuator has been on for a certain time, signalled by an internal timer in the PLC.

The input signal to the PLC may be from manual switches such as the stop/run switch, emergency stop button, reset switches or from mechanically operated switches indicating that a guard is in place, a workpiece is in position or a cylinder has fully stroked.

The PLC is programmed to look for one or more inputs to be present in some logical form to complete a step in the sequence and generate an output signal.

The PLC continuously scans through the program looking for complete statements and then issues an output signal. The PLC program can be shown as a ladder diagram, each rung comprising a logic input statement and an output which may set an internal PLC relay or give an output command to a solenoid. The program can also be written in statement form, the exact format dependent upon the PLC being used, as manufacturers incorporate different facilities and use different commands.

8.2.1 Nomenclature

The following nomenclature is to be used:

A, B, C, etc.   Double-acting pneumatic cylinders
A+, etc.   The extent stroke of cylinder A, etc.
A + p   The pilot signal to cause the A+ stroke of cylinder A
A − p   The pilot signal which initiates the return or retract stroke of cylinder A
A_d The logical complement of A − p, used to reset single-solenoid valves to disable the latch function
V, W, X   Internal PLC relays which can be set on and reset off

V_p Signal which sets relay V
$\overline{V_p}$ Signal which resets relay V
V_d The logical complement of V_p which disables relay V by disabling the latch
T_n Timer number n
C_n Counter number n

The specific addresses used throughout these notes for the programmable logic controller will be as follows:

Inputs   X00, X01, X02, X03, X04, X05, X06, X07, X10, X11, X12, X13, X14, X15, X16, X17

Note: The addresses are to have an octal base, i.e. base 8.

Outputs   Y30, Y31, Y32, Y33, Y34, Y35, Y36, Y37, Y40, Y41, Y42, Y43, Y44, Y45, Y46, Y47

Internal relays: M100, M101, M102, M103, M104, etc.

Timers   T50, T51, etc.

Counters   C60, C61, etc.

The limit switches would be wired up to the input terminals: the common on all the limits to the common on the PLC and the normally open contact on limit switch a_0 to input terminal 10, the normally open contact on limit switch a_1 to input terminal 11, and so on – the output terminals being connected to the valve solenoids.

Example 8.1

Consider the electro-pneumatic circuit shown in Fig. 8.3.

Limit switch a_0 signals that cylinder A is fully retracted and a_1 signals that the extend condition has been reached. Limit switch b_0 signals that cylinder B is fully retracted and b_1 signals that the extend condition has been reached.

The circuit is to operate in the sequence $A+ B+ A- B-$. Determine the command statements for the sequence, assuming a start switch S.

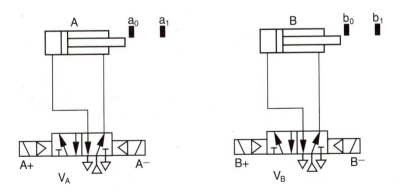

FIGURE 8.3 Electro-pneumatic circuit to provide $A+ B+ A- B-$ using bistable valves.

The $A+$ signal is to occur when the start button is activated and cylinder B is in the retract condition, i.e. limit switch b_0 is operated. Thus

$$A + p = S \text{ AND } b_0$$

$$= S . b_0$$

The $A-$ signal is to occur when cylinder B is fully extended, or

$$A - p = b_1$$

Similarly,

$$B + p = a_1 \quad \text{and} \quad B - p = a_0$$

A ladder diagram can be drawn as shown in Fig. 8.4.

The symbol $\dashv\vdash$ denotes a normally open pair of contacts whereas $\dashv\!\!\!\!/\vdash$ is normally closed. The normally closed contact is the inverse of the normally open contact, therefore $\dashv\vdash$ is the normally open contact of the limit switch and $\dashv\!\!\!\!/\vdash$ is the normally closed contact of the limit switch.

Note: A ladder diagram must always finish with the END rung. The input and outputs are numbered on the ladder diagram. The start signal is connected to input port 04.

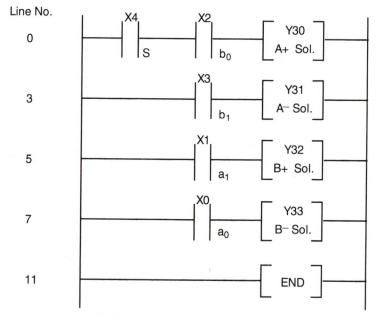

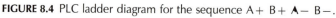

FIGURE 8.4 PLC ladder diagram for the sequence $A+$ $B+$ $A-$ $B-$.

8.2.2 Loading the program into the PLC

A keypad is used to load the program or ladder diagram into the PLC. The keypad has a series of keys so that all the system inputs, outputs, logic functions, relays, counters, timers, etc., can be entered into the PLC. The exact format of the keypad depends upon the PLC used.

The following logic commands are available on most keypads and will be used in the examples in this book.

Keypad	Logic commands
LD	Load into the PLC
LDI	Load inverse into the PLC
AND	AND function
ANI	AND inverse function
ANB	AND block, ANDs block just loaded with the previous block loaded
OR	OR function
ORI	OR inverse function
ORB	OR block, ORs block just loaded with the previous block loaded
OUT	gives an output function
END	program complete

The step instructions can be listed from an inspection of the ladder diagram. Each logic command starts a new step in the instructions; the steps or lines are numbered using an octal numbering system. Consider the ladder diagram shown in Fig. 8.4. The step instructions would be as in Table 8.1, and an explanation of each input or output address can be written alongside if required.

TABLE 8.1 Loading instructions for the sequence A+ B+ A− B−.

Line no.	Input command	Address	Comment
0	LD	X4	Start switch
1	AND	X2	B− Limit switch
2	OUT	Y30	A+ Solenoid
3	LD	X3	B+ Limit switch
4	OUT	Y31	A− Solenoid
5	LD	X1	A+ Limit switch
6	OUT	Y32	B+ Solenoid
7	LD	X0	A− Limit switch
10	OUT	Y33	B− Solenoid
11	END		

8.2.3 Logic block functions ORB and ANB

These commands have to be used in PLC loading instructions when branching occurs in the ladder diagram. When two or more inputs or relays occur in parallel, the second line must be entered into the system using a load command and the ORB command used to link it in parallel, starting at the previous load input command.

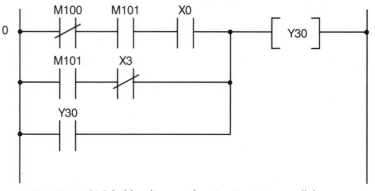

FIGURE 8.5 PLC ladder diagram showing inputs in parallel.

Consider the ladder diagram shown in Fig. 8.5. The loading instructions are given in Table 8.2. When two or more inputs are in series, the second block must be combined with the first block using the ANB command.

TABLE 8.2 Loading instructions for parallel signals or inputs.

Step	Input command	Address	Comments
0	LDI	M100	Internal relay V
1	AND	M101	Internal relay W
2	AND	X0	A− Limit switch
3	LD	M101	Internal relay W
4	ANI	X3	B+ Limit switch
5	ORB		Combines steps 0, 1, 2 in parallel with steps 3, 4
6	OR	Y30	A+ Solenoid
7	OUT	Y30	A+ Solenoid

Consider the ladder diagram in Fig. 8.6. The loading instructions are given in Table 8.3.

TABLE 8.3 Loading instructions for series signals or inputs.

Step	Input command	Address	Comments
10	LDI	M100	Internal relay V
11	AND	M101	Internal relay W
12	LD	X0	A− Limit switch
13	AND	X3	B+ Limit switch
14	OR	Y32	B+ Solenoid
15	ANB		Combines steps 10, 11 in series with steps 12, 13, 14
16	OUT	Y32	B+ Solenoid

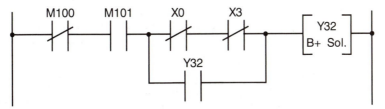

FIGURE 8.6 PLC ladder diagram showing inputs in series.

Example 8.2

Write down the loading instructions for the PLC ladder shown in Fig. 8.7. The loading instructions are given in Table 8.4.

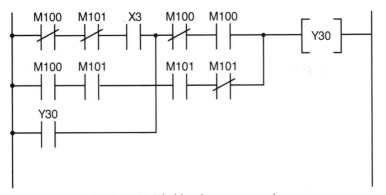

FIGURE 8.7 PLC ladder diagram example.

TABLE 8.4 Loading instructions from the ladder diagram in Fig. 8.7.

Step	Input command	Address	Comments
0	LDI	M100	Internal relay V
1	ANI	M101	Internal relay W
2	AND	X3	B + Limit switch
3	LD	M100	Internal relay V
4	AND	M101	Internal relay W
5	OR	Y30	A + Solenoid
6	ORB		Combines steps 0, 1, 2 with steps 3, 4, 5
7	LDI	M100	Internal relay V
10	ORI	M101	Internal relay W
11	ANB		Combine
12	LD	M100	Internal relay V
13	ORI	M101	Internal relay W
14	ANB		Combine
15	OUT	Y30	A + Solenoid

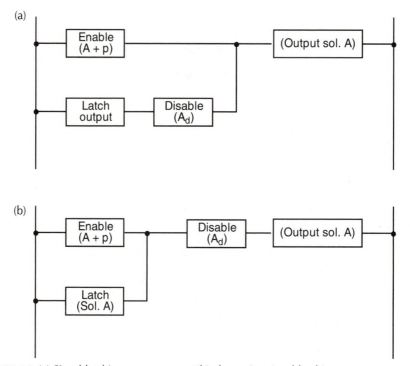

(a)

Enable
(A + p)

(Output sol. A)

Latch
output

Disable
(A$_d$)

(b)

Enable
(A + p)

Disable
(A$_d$)

(Output sol. A)

Latch
(Sol. A)

FIGURE 8.8 (a) Signal-latching arrangement; (b) alternative signal-latching arrangement.

8.2.4 Output signal latching

A single-solenoid valve is a monostable device; when the set signal is removed the valve automatically resets. To keep the valve energised the enable signal has to be latched on; when the signal is removed the latch maintains the output in an energised condition.

A disable signal is used to break the latching circuit. The disable signal is the *inverse* of the ' − ' signal. The two possible ladder rung diagrams are shown in Fig. 8.8 (a); the method shown in Fig. 8.8(b) will be used in the examples in this chapter. For further details see Chapter 6.

Example 8.3

Consider a water tank fitted with two pumps A and B. Pump A pumps water into the tank from an external source and pump B returns the water to the external source. Level switches S1 and S2 are fitted into the tank, as shown in Fig. 8.9.

Pump A is switched on when the fluid level falls below switch S1 and has to be kept on until the level reaches level switch S2. Thus, pump A (see Fig. 8.10) is:

- turned on by S1
- held on by $\overline{S2}$
- turned off by S2

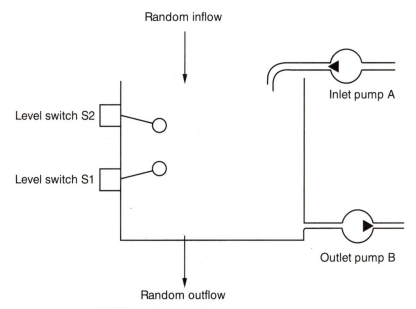

Random inflow

Level switch S2

Level switch S1

Inlet pump A

Outlet pump B

Random outflow

FIGURE 8.9 Pump and water tank arrangement.

FIGURE 8.10 PLC ladder diagram for pump A latching.

Similarly, pump B (see Fig. 8.11) is:

■ turned on by S2

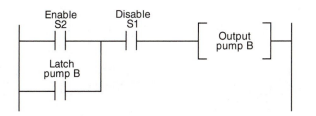

FIGURE 8.11 PLC ladder diagram for pump B latching.

- held on by S1
- turned off by $\overline{S1}$

For pump A, the $A - p = S2$, whereas the hold-on or turn-off signal is $\overline{S2}$. This is also the disable signal for the latch circuit. Thus

$$A_d = \overline{A - p}$$

or the A disable pilot is the inverse of the $A-$ pilot.

A similar system applies to the disabling of pump B, where

$$B - p = \overline{S1} \quad \text{and} \quad B_d = S1 = \overline{B - p}$$

(*Note*: In this example one pump will be running continuously.)

Example 8.4

For the electro-pneumatic circuit shown in Fig. 8.12 using single-solenoid valves, draw the ladder diagram for the sequence $A+ \, B+ \, A- \, B-$.

Note: The valves are monostable and the solenoid signal has to be latched.

Solution

Use a start signal S. By inspection

$$A + p = S . b_0$$

$$A - p = b_1$$

A_d, the disable signal, is:

$$A_d = \overline{A - p} = \overline{b}_1$$

From example 8.1:

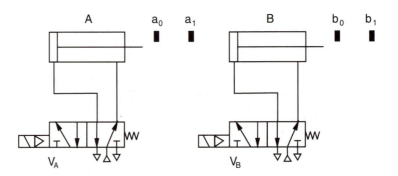

FIGURE 8.12 Electro-pneumatic circuit to provide $A+ \, B+ \, A- \, B-$ using monostable valves.

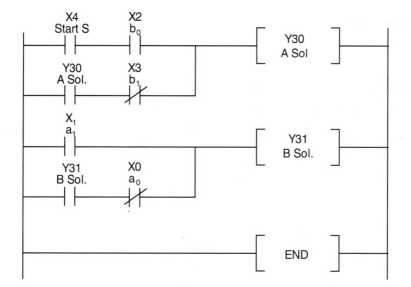

FIGURE 8.13 PLC ladder diagram to provide $A+$ $B+$ $A-$ $B-$ using monostable valves.

TABLE 8.5 Loading instructions for the sequence $A+$ $B+$ $A-$ $B-$ using monostable valves.

Step	Input command	Address	Comments
0	LD	X4	Start Switch
1	AND	X2	$B-$ Limit Switch
2	LD	Y30	A Solenoid
3	ANI	X3	$B+$ Limit Switch
4	ORB		
5	OUT	Y30	A Solenoid
6	LD	X1	$A+$ Limit Switch
7	LD	Y31	B Solenoid
10	ANI	X0	$A-$ Limit Switch
11	ORB		
12	OUT	Y31	B Solenoid
13	END		

$$B + p = a_1$$

$$B - p = a_0$$

and

$$B \text{ disable} = B_d = \overline{B - p} = \bar{a}_0$$

The ladder diagram and step instructions are shown in Fig. 8.13 and Table 8.5 respectively.

Example 8.5

The electro-pneumatic circuit shown in Fig. 8.12 is to operate in the sequence A+ B+ B− A−. Draw the ladder diagram for a PLC and list the step instructions.

Solution

Consider the sequence to be a cascade. Divide it into groups as a cascade sequence.

$$A+ \; B+ \; | \; B- \; A-$$
$$\text{Group I} \; | \; \text{Group II}$$

Use an internal PLC relay W to differentiate between the groups, i.e.

$$\text{Group I} = \bar{W}$$
$$\text{Group II} = W$$

Always commence the sequence with all the relays de-energised.

Next determine the command statements for each step, using an on/off switch S. By inspection of the sequence write down the commands for each step.

$A + p = S \cdot \bar{W}$ the first operation group I
$B + p = \bar{W} \cdot a_1$
$B - p = W$ the first operation in group II
$A - p = W \cdot b_0$

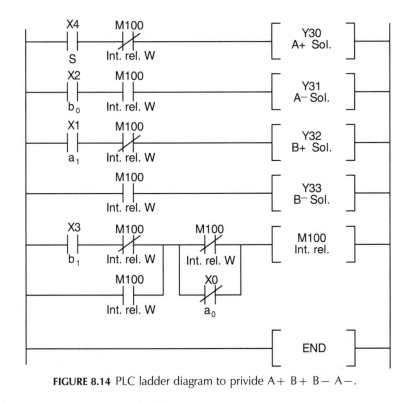

FIGURE 8.14 PLC ladder diagram to privide A+ B+ B− A−.

Command statements for the pilots are

$$W_p = \bar{W} \cdot b_1$$
$$\bar{W}_p = W \cdot a_0$$

If the internal relays of the PLC have a set/reset capability, the expressions for W_p and W_d can be used directly. If not, then W_d has to be found:

$$W_d = \bar{W}_p = \overline{W \cdot a_0} = \bar{W} + \overline{a_0}$$

The ladder diagram in Fig. 8.14 assumes that the internal relay has to be latched. The step input instructions are given in Table 8.6.

TABLE 8.6 Loading instructions for the sequence $A+ B+ B- A-$.

Step	Input command	Address	Comments
0	LD	X4	Start switch
1	ANI	M100	Internal relay
2	OUT	Y30	A+ Solenoid
3	LD	X2	B− Limit switch
4	AND	M100	Internal relay
5	OUT	Y31	A− Solenoid
6	LD	X1	A+ Limit switch
7	ANI	M100	Internal relay
10	OUT	Y32	B+ Solenoid
11	LD	M100	Internal relay
12	OUT	Y33	B− Solenoid
13	LD	X3	B+ Limit switch
14	ANI	M100	Internal relay
15	LD	M100	Internal relay
16	ORB		
17	LDI	M100	Internal relay
20	ORI	X0	A− Limit switch
21	ANB		
22	OUT	M100	Internal relay
23	END		

8.2.5 Multiple operations

When a cylinder operates non-repetitively more than once in a sequence this can be taken care of by writing down the command statement for each operation and incorporating these in the ladder diagram.

Example 8.6

Consider the electro-pneumatic circuit in Fig. 8.12 and obtain the step ladder diagram for the sequence

$$A+ B+ A- B- A+ A-$$

Treat the sequence as a cascade circuit.

The first step is to divide the sequence into groups and allocate relay states to each group. Note the first group in the sequence must have all internal relays de-energised. The relay grouping is as follows:

1. Relay W   Groups are $\bar{W}$ W
2. Relays V, W   Groups are $\bar{V}\bar{W}$ $\bar{V}W$ VW $V\bar{W}$
3. Relays U, V, W Groups can be

$$\bar{U}\bar{V}\bar{W} \quad \bar{U}\bar{V}W \quad \bar{U}VW \quad UVN$$
$$U\bar{V}W \quad U\bar{V}\bar{W} \quad UV\bar{W} \quad \bar{U}V\bar{W}$$

giving eight groups, or

$$\bar{U}\bar{V}\bar{W} \quad U\bar{V}\bar{W} \quad \bar{U}VW \quad UVW$$
$$U\bar{V}W \quad U\bar{V}\bar{W}$$

giving six groups.

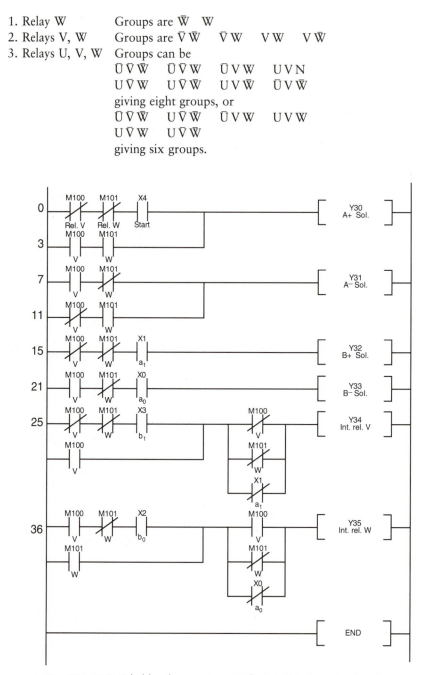

FIGURE 8.15 PLC ladder diagram to provide $A+$ $B+$ $A-$ $B-$ $A+$ $A-$.

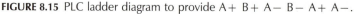

The sequence given may be split into groups as follows:

$$\begin{array}{c|c|c|c}
A+\ B+ & A-\ B- & A+ & A- \\
I & II & III & IV \\
\bar{V}\,\bar{W} & V\,\bar{W} & V\,W & \bar{V}\,W
\end{array}$$

The ladder diagram in Fig. 8.15 is drawn assuming the internal relays have to be latched.

The command statements will be

$$A + p = S \cdot \overline{VW} + VW \quad \ldots A+ \text{ occurs twice in the sequence.}$$
$$A - p = V\bar{W} + \bar{V}W \quad\quad \ldots A- \text{ occurs twice in the sequence.}$$
$$B + p = \overline{VW} \cdot a_1$$
$$B - p = V\bar{W} \cdot a_0$$
$$V_p = \overline{VW} \cdot b_1$$
$$\bar{V}_p = VW \cdot a_1$$
$$V_d = \overline{VWa_1} = \bar{V} + \bar{W} + \overline{a_1}$$
$$\bar{W}_p = VW \cdot b_0$$
$$\overline{W}_p = \bar{V}W \cdot a_0$$
$$W_d = \overline{\bar{V} \cdot W \cdot a_0} = V + \bar{W} + \overline{a_0}$$

The step input instructions for ladder diagram Fig. 8.15 are given in Table 8.7.

TABLE 8.7 Loading instructions for the sequence $A+\ B+\ A-\ B-\ A+\ A-$.

Step	Command and address	Comment	Step	Command and address	Comment
0	LDI M100	Rel. V	25	LDI M100	Rel. V
1	ANI M101	Rel. W	26	ANI M101	Rel. W.
2	AND X4	Start sw.	27	AND X3	B+ Lt sw.
3	LD M100	Rel. V	30	OR M100	Rel. V
4	AND M101	Rel. W	31	LDI M100	Rel. V
5	ORB		32	ORI M101	Rel. W
6	OUT Y30	A+ Sol.	33	ORI X1	A+ Lt sw.
7	LD M100	Rel. V	34	ANB	
10	ANI M101	Rel. W	35	OUT Y34	Rel. V
11	LDI M100	Rel. V	36	LD M100	Rel. V
12	AND M101	Rel. W	37	ANI M101	Rel. W
13	ORB		40	AND X2	B− Lt sw.
14	OUT Y31	A+ Sol.	41	OR M101	Rel. W
15	LD M100	Rel. V	42	LD M100	Rel. V
16	ANI M101	Rel. W	43	ORI M101	Rel. W
17	AND X1	A+ Lt sw.	44	ORI X0	A− Lt sw.
20	OUT Y32	B+ Sol	45	ANB	
21	LD M100	Rel. V	46	OUT Y35	Rel. W
22	ANI M101	Rel. W	47	END	
23	AND X0	A− L. sw.			
24	OUT Y3				

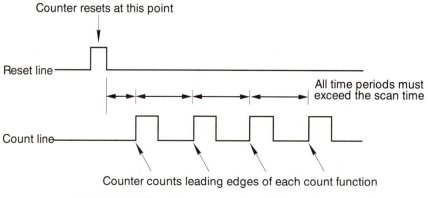

FIGURE 8.16 Schematic representation of the counter operation.

8.2.6 Repetitive operations (counters)

When an operation is to be repeated several times consecutively, a counter on board the PLC is used. The counter has two input lines, one to reset the counter to the start condition and restore the count valve, the second input is the function to be counted. The counter resets with a rising or positive going pulse on the reset line; and this pulse must be removed or low during the count operation or the counter will reset. The count line input has to be a rising or positive pulse for each function to be counted. The time of each pulse plus the time between pulses must be greater than the time taken by the PLC to do a complete scan through its program. This scan time will depend upon the length of the program and the type of PLC used. The counter operation is shown schematically in Fig. 8.16. The number of counts is fed into the PLC from the keypad using a count key.

When the counter is in the SET or RESET condition it is in the unoperated condition, and remains in this condition until it has completed its count when it returns to the operated condition. When the counter is reset the normally closed contacts are made.

Example 8.7

Referring to Fig. 8.12, draw a PLC ladder diagram for the sequence

$$A+ B+ B- B+ B- B+ B- A-$$

If a cascade system were used it would necessitate six groups, but by using a counter this can be reduced to two groups. The sequence can be written as:

$$A+ \left(\underset{\times 3}{B+ \ B-} \right) A-$$

This will give two groups as the repeating function can be considered as a single operation.

$A+ [(B+ B-) \times 3]$	$A-$
Group I	Group II
Relay $\overline{W}$	Relay W

As single-solenoid valves are used, the signals to the solenoids have to be latched. By inspection, the command statements will be

$$A + p = \overline{W} \cdot S \quad \text{where S is the start signal}$$
$$A - p = W$$

thus

$$A_d = \overline{A - p} = \overline{W}$$
$$B + p = \overline{W} \cdot a_1 \cdot b_0 \cdot \overline{C60}$$

(*Note*: As B+ is controlled by a count function, B must not start to extend until it has fully retracted.

$$B - p = \overline{W} \cdot b_1$$

So

$$B_d = \overline{\overline{W} \cdot b_1} = W + \overline{b_1}$$
$$W_p = \overline{W} \cdot b_0 \cdot C60$$

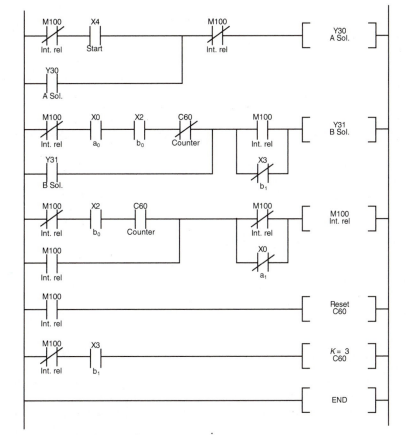

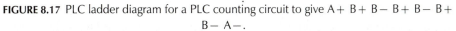

FIGURE 8.17 PLC ladder diagram for a PLC counting circuit to give A+ B+ B− B+ B− B+ B− A−.

$$\overline{W}_p = W . a_0$$

$$W_d = \overline{W . a_0} = \overline{W} + \overline{a_0}$$

The counter must be reset after the count is complete. In this case, at the start of group II,

$$\text{Counter reset CR} = W$$

The count function is the first step in the operation to be counted, i.e. $B+$. The count constant K is the number of operations required. Thus,

$$\text{Counter function CF} = \overline{W} . b_1 \quad (K = 3)$$

A ladder diagram can now be constructed, as shown in Fig. 8.17.

The command statement for ladder diagram 8.17 will be given in Table 8.8.

TABLE 8.8 Loading instructions for the sequence $A+ \ B+ \ B- \ B+ \ B- \ B+ \ B- \ A-$.

Step	Command and address	Comment	Step	Command and address	Comment
0	LDI M100	Int. rel.	17	LDI M100	Int. rel.
1	AND X4	START	20	AND X2	B− Lt sw.
2	OR Y30	A+ Sol.	21	AND C60	COUNTER
3	LDI M100	Int. rel.	22	OR M100	Int. rel.
4	ANB		23	LDI M100	Int. rel.
5	OUT Y30	A+ Sol.	24	ORI X0	A− Lt sw.
6	LDI M100	Int. rel.	25	ANB	
7	AND X0	A− Lt sw.	26	OUT M100	Int. rel.
10	AND X2	B− Lt sw.	27	LD M100	Int. rel.
11	ANI C60	Counter	30	RST C60	COUNTER
12	OR Y31	B+ Sol.	31	LD M100	Int. rel.
13	LD M100	Int. rel.	32	AND X3	B+ Lt sw.
14	ORI X3	B+ Lt sw.	33	OUT C60	COUNT
15	ANB			K3	FUNCTION
16	OUT Y31	B− Sol.	34	END	

8.2.7 Timers

Timers built into the PLC may be used to delay or to time an operation; for example, to retract a cylinder after a certain time delay or to keep a motor operating for a time period. The PLC 'on board' timers need an input line which goes high to start the timer and must stay high during the timer period. The timer output goes high after the timer period and remains high after the timer resets. The time delay period is usually adjustable in 0.1 s intervals. The operation of a timer is shown diagrammatically in Fig. 8.18.

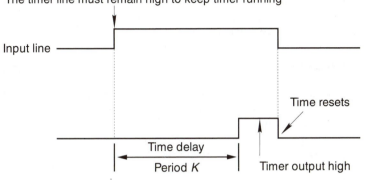

FIGURE 8.18 Schematic representation of the timer operation.

Example 8.8

Consider the pneumatic circuit shown in Fig. 8.3. Because of restrictions in the system it is impossible to fit a limit switch to indicate the extended condition of cylinder B. The machine sequence is to be

$$A+ \ A- \ B+ \ B-$$

As there is no b_1 limit switch, a time delay is to be used to initiate $B-$. Construct the ladder diagram and write down the output commands for a PLC to give this sequence.

Solution

Use a cascade solution (where W is an interval relay):

$$
\begin{array}{c|c|c}
A+ & A- \ B+ & B- \\
\text{Group I} & \text{Group II} & \text{Group I} \\
\overline{W} & W & \overline{W}
\end{array}
$$

$$\overline{A} + p = W \cdot S \cdot b_0$$

$$A - p = W$$

$$B + p = W \cdot a_0$$

$$\overline{B} - p = \overline{W}$$

$$\overline{W}_p = \overline{W} \cdot a_1$$

$$\overline{W_p} = W \cdot T50$$

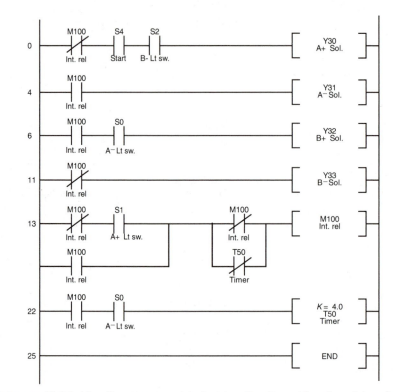

FIGURE 8.19 PLC ladder diagram to provide A+ A− B+ B− with a time delay after B+.

therefore

$$W_d = \overline{W \cdot \overline{T50}} = \overline{W} + T50$$

Timer enable (input): This goes high to show the time delay. Let the timer enable be set by the A− signal in group II.

Assume that the extend stroke of cylinder B takes 3 s, then set the time delay to, say, 4 s. This can be adjusted when the machine has been run. Thus,

$$\text{T enable} = W \cdot a_0 \quad (K = 4.0 \text{ s})$$

The ladder diagram for these expressions is shown in Fig 8.19. The command statements for the ladder diagram in Fig. 8.19 are given in Table 8.9.

TABLE 8.9 Loading instructions for the sequence A+ A− B+ delay B−.

Step	Command and address	Comment	Step	Command and address	Comment
0	LDI M100	Int. rel.	12	OUT Y33	B − Sol.
1	AND S4	START	13	LDI M100	Int. rel.
2	AND S2	B − Lt sw.	14	AND S1	A+ Lt sw.
3	OUT Y30	A+ Sol.	15	OR M100	Int. rel.
4	LD M100	Int. rel.	16	LDI M100	Int. rel.
5	OUT Y31	A− Sol.	17	ORI T50	TIMER
6	LD M100	Int. rel.	20	ANB	
7	AND S0	A− Lt sw.	21	OUT M100	Int. rel.
10	OUT Y32	A− Lt sw.	22	OUT M100	Int. rel.
11	LDI M100	Int. rel.	23	AND S0	A− Lt sw.
			24	OUT T50	TIMER
				$K =$ 4.0	Time delay 4.0 secs
			25	END	

8.3 Fieldbus systems

Within modern plant there has developed a requirement for the centralised control of multi-functional systems. Where a plant has a number of different machines or an extensive machine with actuators positioned in remote groups, a centralised computer or PLC provides integrated control, centralised collection of data, monitoring and simplified machine process modifications.

'Fieldbus' systems have been designed that will allow a centralised PLC or computer to control remote devices such as solenoid valves, sensors, etc. Communication is through a serial two-wire bus, linking through a series of slave devices in the system. Figure 8.20 shows a typical arrangement for a fieldbus control system. The two-wire system links in and out of each fieldbus interface until each module is connected.

Each machine or part of a machine can have its actuators driven by local valve islands or manifolds containing the solenoid valves. The valve islands or manifolds can have their own fieldbus interface output module which can be connected locally to a fieldbus interface input module. The serial communication from the master PLC is passed along the two-wire bus until it locates its specific address, where instructions will be given to energise or de-energise the desired combination of solenoid valves. The resulting actuator movements will be identified by the switches and/or sensors and fed back to the fieldbus input module and back through the serial link to the master controller.

It will be appreciated that the wiring requirements of a fieldbus system are much simpler than conventional systems.

At the time of writing there are a number of different systems currently operating in industry and more are being developed. These can be split into two categories: open systems and proprietary systems.

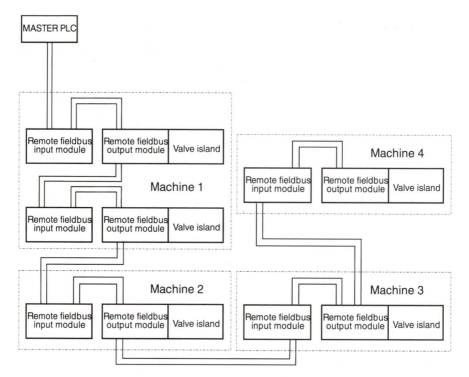

FIGURE 8.20 A typical layout for a fieldbus control system.

An open system is one which is manufactured to a particular standard. This means that any manufacturer's equipment with a serial interface conforming to that standard will be able to communicate with any other manufacturer's equipment with a similar interface.

A proprietary system is one in which only the system manufacturer's equipment is able to communicate. Although this is very limiting, the systems are usually quite user friendly.

CHAPTER 9

Maintenance

9.1 Component maintenance

Prior to carrying out any stripping of components it is advisable to have the following:

1. A clean work area.
2. An exploded view or sectional drawing of the component.
3. Any special tools required to dismantle/rebuild the component.
4. A set of spare seals.

When taking a component down, place any parts removed on a clean surface in the order in which they were removed. If spares have to be ordered before a component can be repaired, the component must be re–assembled to await repair.

9.1.1 Valve seals

These are often specially moulded for the valve manufacturer and even 'O' ring seals may be to closer limits than those given in British Standards. Consequently, only seals obtained from the valve manufacturer should be used. Various elastomers are used in the manufacture of seals, dependent upon the application. If a special fluid is used or the operating temperature is outside the normal limits (0–100 °C) special seals may have to be used.

When an incompatible fluid is used on a seal, the usual result is swelling and softening of the elastomer. Some cleaning agents such as trichloroethylene cause swelling of all elastomers and should not be used in any circumstances. Standard seals are compatible with air, water, mineral-based oils (including petrol, diesel and paraffin), but not castor-based oils, acids and alkalis.

9.1.2 Valve maintenance

Although routine maintenance of all pneumatic components could reduce breakdown time, it is very seldom carried out. This is due to the large number of components involved and the amount of time required. It is recommended, however, that filters, lubricators and regulators be regularly serviced, together with any valves which only operate in emergency conditions. It is possible for a valve to stick due to a buildup of deposits if it is not used regularly.

The basic components of a 3-port, 2-position, normally closed, plunger-operated, spring return poppet valve are shown in Fig. 9.1. This valve body is used for other valves, hence the additional plugged port.

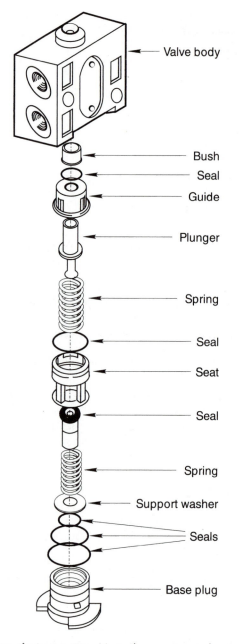

Valve body

Bush

Seal

Guide

Plunger

Spring

Seal

Seat

Seal

Spring

Support washer

Seals

Base plug

FIGURE 9.1 Exploded view of a 3-port, 2-position, plunger-operated spring return poppet valve.

The basic parts of a lever-operated 5-port spool valve are depicted in Fig. 9.2. The seals and valve spool are available as a wearing parts kit.

When a number of one particular type of valve are used in a plant, a spare valve should always be carried. If there are only one or two valves of a particular type, spare

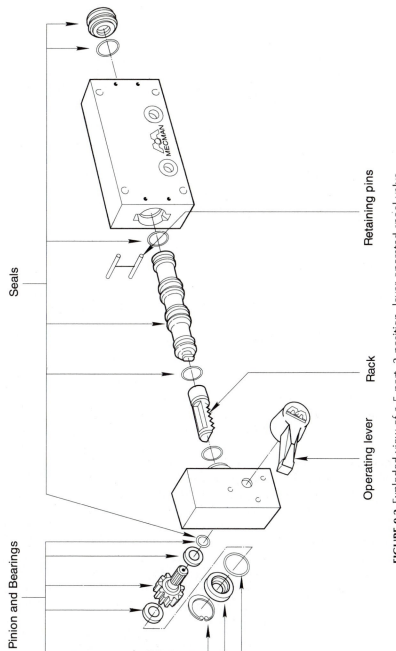

Seals

Retaining pins

Rack

Operating lever

Pinion and Bearings

FIGURE 9.2 Exploded view of a 5-port, 2-position, lever-operated spool valve.

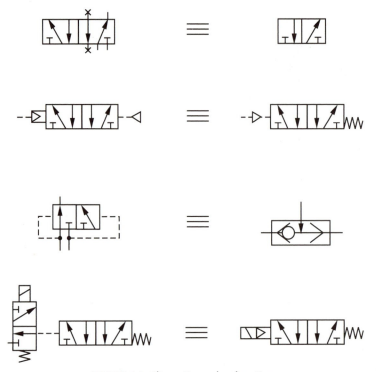

FIGURE 9.3 Alternative valve functions.

seals and springs should be kept in stock. Even where there is a pneumatic supplier in the area there may be a few days delay in obtaining a replacement valve. Often slight modifications to a valve can alter its operation, enabling one valve to be modified to replace a valve of a different type. Some alternative valve functions are shown in Fig. 9.3.

It can be seen from the foregoing that, in an emergency, it is possible to replace failed valves with alternative valves modified to give the required function. The correct valve should be replaced in the system at the earliest possible opportunity.

When repairing a valve cleanliness is of paramount importance, as dirt in a valve is one of the main causes of malfunction. The valve components can be cleaned with paraffin; in some cases it may be necessary to use fine emery cloth or metal polish, but the component must be thoroughly cleaned afterwards. When replacing 'O' ring seals care must be taken to ensure that the seal is not twisted, otherwise it may leak. When a seal has to pass over a spool containing a number of grooves, a tubular tool fitting over the spool will be found useful. This prevents damage to the 'O' ring. A smear of lubricating oil on the seals will help fitting. After final assembly the component should be lightly oiled and fully tested before being replaced in the circuit.

9.1.3 Solenoid valve maintenance

If possible the operating voltage of all the solenoid valves throughout the factory should be standardised. When there are various voltages, care must be taken when replacing solenoids to ensure that the operating voltage is correct. It is advisable to keep spare solenoids in stock, especially if they are of an unusual voltage or specification.

In some types of solenoid valves a separate air supply is required to the solenoid section. Another valve which appears to be identical may have the pilot supply to the solenoid internally piped. Consequently, when changing valves it is very important to check the part number. When replacing an air-operated both ways, an air reset valve or a double-solenoid valve, ensure that the spool is in the correct position – i.e. the correct ports internally connected within the valve. In some circuits, a reversed spool may stop the circuit operating altogether; in others it may cause the circuit to start at some intermediate point in the cycle with consequential damage.

9.1.4 Cylinder construction

The detailed construction of a pneumatic cylinder varies from manufacturer to manufacturer. The components of a basic double-acting cylinder are shown in Fig. 9.4. The piston and piston rod are connected to each other in one of the following ways:

1. By a resin-type adhesive.
2. The piston being in halves and bolted together over a collar on the piston rod.
3. Screwing the piston onto rod plus locknut, or alternative locking methods.
4. Grooved piston rod and circlips.
5. One-piece piston and piston rod – seldom used in modern construction.

The end caps are secured to the piston body by screws, threading the body, or the use of tie-rods. Whatever the type of construction, a sectional drawing of the cylinder and parts list is very helpful when carrying out cylinder repair.

9.1.5 Cylinder seals

Only spare seals supplied by the cylinder manufacturer should be used. Generally the seals are suitable for operation up to 150°C, dependent on cylinder type and make. Some manufacturers offer cylinders with special seals for operation at temperatures up to 300°C; otherwise all comments made under the section on valve seals apply.

9.1.6 Cylinder installation

Great care must be taken to ensure that no side load can occur on the piston rod, otherwise excessive seal wear will occur. The cylinder securing point must be strong enough to withstand the fully working load. If trunnion or swivel mountings are used on the cylinder, flexible tubing must be used. The pipe used must be large enough to pass the full flow of air with negligible pressure drop. Sharp bends in the lines should be avoided.

9.1.7 Cylinder maintenance

To give long cylinder life, it is essential to exclude all dirt and moisture from the cylinder and to ensure that the air supply is adequately lubricated. Where the cylinder is operating in dirty or corrosive conditions a gaiter should be used to protect the cylinder rod (see Fig. 9.5).

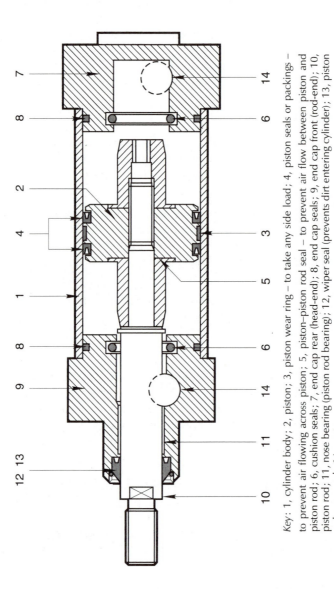

Key: 1, cylinder body; 2, piston; 3, piston wear ring – to take any side load; 4, piston seals or packings – to prevent air flowing across piston; 5, piston–piston rod seal – to prevent air flow between piston and piston rod; 6, cushion seals; 7, end cap rear (head-end); 8, end cap seals; 9, end cap front (rod-end); 10, piston rod; 11, nose bearing (piston rod bearing); 12, wiper seal (prevents dirt entering cylinder); 13, piston rod nose seal; 14, ports.

FIGURE 9.4 Component detail for a double-acting cylinder.

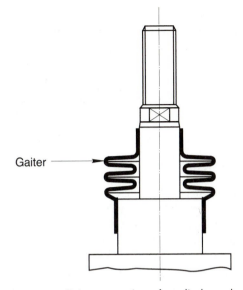

FIGURE 9.5 Gaiter protection of a cylinder rod.

The cylinder end caps should be removed at regular intervals, and the piston assembly withdrawn, cleaned and inspected.

The rod assembly and cylinder should be cleaned by washing in paraffin. **A degreasing fluid must not be used**.

When assembling the piston into the cylinder, care must be taken not to damage the seals on the piston. Placing the cylinder body in a vice will leave both hands free to guide the piston into the body.

If piston rod seals have to be replaced, the special tools available from the manufacturer should be used.

These tools fit over the threaded end of the piston, allowing the seal to be pushed down over it. The seals should be lubricated before fitting.

9.2 Installation of pneumatic equipment

9.2.1 Installing a complete machine

If a complete new machine is to be installed, usually only the service connections have to be made. The air supply must be adequate for the machine, and if the machine is not fitted with a filter, regulator and lubricator, these must be fitted to the supply if required, together with an air shut-off valve. The lubricator should be filled with the correct grade of lubricating oil.

Note: Check that a lubricator is needed, as more and more use is being made of components that do not need lubricating.

Before switching on, check that all pipes are connected, that cylinders are in their start position and are correctly secured, that trip valves are correctly positioned, and that all loose parts, tools, etc., are removed from the machine.

Switch on the air supply, adjust the pressure regulator to the recommended working pressure and then start the machine. If the machine has dual manual/auto controls, start initially with the manual control. Adjust lubricators as necessary to suit the circuit requirements.

9.2.2 Building a new circuit

Equipment supplied from pneumatic component manufacturers is already lubricated; there is no need for any further lubrication during assembly. Components are sealed with plastic dust caps or shrink wrapped, neither of which should be removed until the pipe connections are to be made. Any protective coverings, such as plastic film on piston rods, should not be removed until the last possible moment.

A test air line should be available for testing each part of the circuit as it is built. The test air line should have a built-in filter, regulator and, where required, lubricator and a stop valve or blow gun outlet.

Cylinders should be installed first, and these must be checked for alignment to ensure that there are no lateral forces. Dial gauges can be used for this purpose. If the cylinder is used in a clamping application, the clamp pad must be square to the piston rod. If a trunnion or pivot mounted cylinder is used, this must oscillate freely, the pivot pins being lubricated prior to assembly.

When a cylinder is correctly aligned the movement will be smooth and uniform. With small cylinders and light loads it is possible to test the movement manually; for large cylinders the test air line can be used with a suitable flow regulator. The cushions should not be adjusted until the circuit is fully piped up, and then adjusted when operating under full load conditions.

When building up the circuit, position all mechanical valves, i.e. trip valves, and check their correct operation. Then position all manual valves, making sure that they are accessible to the operator.

Next position the filter, regulator, lubricator assembly and the main isolating valve, which again must all be accessible for servicing. The other valves can now be positioned either in a control box or on the machine. The cylinder direction control valves should be as near to the cylinder as possible in order to minimise air consumption and to decrease the response time.

The circuit can now be piped up. It is advisable to deal with one pipe run at a time wherever possible. On complex circuits the use of colour-coded pipes simplifies circuit tracing. When piping up a double-pilot-operated valve, use the test air line to pilot the valve into its start condition. As each section of the circuit is built the test air line can be used to check for correct operation and 'air leaks'. As a pipe is connected it can be marked off on the circuit diagram. When the circuit is completed, set to manual operation and check, adjusting all flow control valves, pressure sensitive valves, etc., before changing to automatic operation. If there is no manual control, all the settings of the valves will have to be completed when the circuit is in 'automatic cycling'.

The usual fault occurring on wiring up a circuit is for a cylinder to stroke the wrong way, due to a direction control valve being piped incorrectly. This will be avoided if the test air line is used at each stage as suggested.

When the system has been run 'dry' it should be tested under working conditions, when it may be necessary to adjust some of the speed controls, etc. Some pneumatic systems operate extremely rapidly, and, consequently, at normal speeds it is impossible to visually check the operating sequence. Testing then involves the use of either an electrical or electronic monitoring system, i.e. a stroboscopic flash unit, or the use of exhaust restrictors to slow the action of the cylinders to a suitable speed for visual checking.

If a circuit has been tested step by step as it was built up, the complete circuit should work correctly. The circuit may work 'dry' but not under load; this could indicate undersized components such as cylinders, motors, etc.

If a fault is found as one of the steps of assembly is completed and this appears to be a design fault, the designer must be notified. Any modifications made to the circuit sequence, components, types or sizes must be recorded on the drawings. All obsolete drawings must be destroyed.

Before a system is put into operation, the following series of safety checks must be made:

1. The effect of an air supply failure at any point in the cycle.
2. The effect of variations in supply pressure.
3. The effect of a cylinder jamming – i.e. not fully stroking.
4. The effect of incorrect operation of manual control valves.
5. All guards and safety devices are in place and operative.

If the process controlled by the pneumatic circuit is such that a costly product would be ruined, or the machine damaged by non-completion of a cycle due to an air supply failure, an air receiver can be inserted in the supply line immediately prior to the circuit. The air receiver must be of sufficient capacity to hold enough air for one complete cycle of the machine.

Test indicators

Small test indicators, which are single-acting miniature cylinders, are sometimes built into a pneumatic circuit to monitor the circuit operation. These indicators can be very useful in fault finding, but if they are continuously 'on line' they may fail, leading to erroneous conclusions.

9.3 Fault finding

9.3.1 Safety rules

1. Before working on any circuit lower off or chock up any cylinder which could allow a load to fall.
2. Ensure that no unauthorised person can switch air on or off the system.
3. In no circumstances block an air line by hand as air can penetrate the skin. Use a stop valve to switch air on or off.

4. Loose air lines can fly about under air pressure. It is sufficient just to crack a joint to determine if a line is pressurised.
5. If an electrical fault is suspected, a qualified electrician should check the electrical systems.
6. In no circumstances must mechanical trip valves be operated by hand. A screwdriver blade, spanner, etc., may be used.

9.3.2 Fault-finding procedure: production machines

Information required about machine

It is essential that an up-to-date circuit diagram is available for each pneumatic system. Whenever any modification is made to a circuit, all drawing must be altered or replaced with up-to-date copies.

The circuit diagram and/or the operating instructions should contain the following information:

1. Full details of each component, i.e. for a cylinder: bore, stroke, type of mounting cushioning, manufacturer and part number. Each component should be numbered on the circuit diagram and a corresponding number painted or otherwise shown on the actual component.
2. Minimum operating pressure of system.
3. Operating sequence. If there are various sequences these must all be stated.
4. Setting of pressure regulators used in the system.
5. Starting and stopping procedure. Position of manual valves.

Information required from operator concerning breakdown

1. Type of fault. Has machine stopped partway through cycle?
2. Did fault occur suddenly or has it developed over a period of time?
3. Have any unauthorised modifications been carried out on the machine?
4. Has component been altered (altering shape could prevent trip valves, etc., operating correctly)?
5. Has operator attempted to correct fault? If so, anything that has been altered must be returned to original position.

Initial check list

1. Air supply to machine is turned on and set to correct pressure. (*Note*: 'Shut off' valves do not always give a perfect seal, but allow slight leakage. This can result in pressure gauges showing correct pressure even when 'shut off' valve is closed, and machine could complete part of a cycle on the small amount of air trapped in the system between the machine and the shut-off valve.) Ensure that the supply line is not trapped.
2. Cylinders are in correct position to operate appropriate trips. Trip valves may have become loose on mountings, or cylinder may not have fully completed a stroke.
3. Manual valves in correct position and any pressure regulators at correct setting.

9.3.3 Logical fault finding

After checking correct supply and setting of valves, if a fault still exists determine the point in the cycle at which the sequence stops. (Note the position of each cylinder and compare with the required sequence.)

Prepare a list of all the possible causes of the fault and against each one identify the checks and tests you will make. Prioritise the list, then commence the checking process following the list meticulously to ensure that no possible problems are overlooked.

When the fault has been found and rectified, consider its causes and consequences. Has it put the machine out of sequence? Is the fault likely to occur again? etc. Enter a detailed report into the machine maintenance log to aid future fault finding and identify any machine weaknesses.

Example 9.1

Consider the circuit shown in Fig. 9.6. The machine stops with cylinder B fully retracted and cylinder A fully extended. The fault suddenly occurred, nothing on the machine has been altered.

Check: Air supply correct.

Determine the point in the cycle where the machine has stopped.

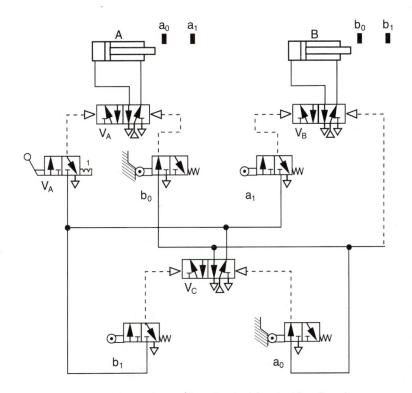

FIGURE 9.6 Pneumatic machine circuit giving $A+$ $B+$ $B-$ $A-$.

Solution

This can be either after A+ and before B+ or after B− and before A−. If the point cannot be decided by visual inspection determine – by checking the outlet ports of V1 (the group control valve) – if air is being supplied to group I or group II.

Let us assume that the fault occurs between A+ and B+. This means that cylinder B has not extended. A series of checks can be carried out, as shown in Table 9.1, to help locate the fault.

TABLE 9.1 Checks and actions taken to determine the cause of failure if cylinder B does not extend.

Check	Result	Action
Air on cylinder B extend port	Air present	Cylinder jammed: release air on retract port Exhaust restrictor blocked Failure of valve V_B
	No air	Proceed to next check
Air on V_B+ pilot port	Air present	Air on both pilot ports of V_B locking spool Failure of V1 (both groups supplied with air) Air on V_B+ port only – failure of V_B.
	No air	Proceed to next check
Air in outlet of a_1	Air present	Line V_B blocked
	No air	Check supply to a_1. If correct supply and valve operated, the valve must have failed

Example 9.2

Consider the pneumatic cascade circuit shown in Fig. 9.7. This is a two-group cascade circuit with time delays and pressure sensing.

Before trying to determine the possible cause of the fault, study the circuit diagram thoroughly, going through its operation step by step. Make sure that the function of each component is understood.

When trying to determine the fault read the symptoms carefully, then consider the effect of maladjusted or malfunctioning components. It is possible that there is more than one fault to give the symptoms, in which case list all the possibilities. In practical fault finding, when there is more than one possible fault, always check those faults that do not require any dismantling of the circuit first, e.g. setting of flow controls, pressure switch settings, operation of trip valves and so on. The symptoms of a series of possible faults are shown in Table 9.2 together with the possible causes of these faults.

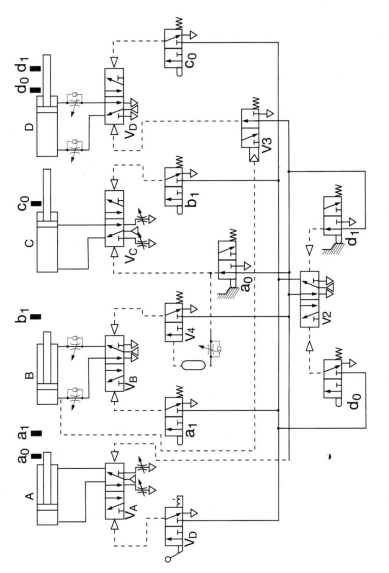

FIGURE 9.7 Pneumatic machine circuit giving A+ B+ C− D− A− C+ B− D+.

TABLE 9.2 Symptoms and possible causes of a series of faults for the circuit shown in Fig. 9.8.

Symptom	Possible cause
1. Sequence correct but time delay before B— is too short. Adjustment of time delay flow control valve has no effect	Time delay check valve leaking. Flow control valve not working. (Needle not being driven by control knob.) Trip valve a_1 not fully exhausting previous signal
2. Sequence stops with cylinders A, B and D extended, cylinder C retracted	Trip valve c_0 not operating correctly. Valve V_D jammed.
3. After a shut-down period the machine will not start	Air supply pressure too low Valve V2 in an intermediate position. Air trapped on both sides of valve V_A. Valve V_A jammed. No output from d_2.
4. Cylinder D moves normally but air bleeds from the exhaust port of V_D when the cylinder is fully extended	Seal failure on cylinder D. Internal leakage in valve V_D
5. Cylinder B is jerky on the extend stroke	Low air pressure Excessive load on the cylinder Cylinder guides or load guides require lubrication
6. Machine stops with cylinders B and C extended and A and D retracted.	Valve V4 jammed Time delay restrictor fully wound in or blocked Valve V_B jammed

Once the list of all possible causes of the failure is complete, then a checking priority can be established. If one now follows the list of prioritised checks, no apparently trivial fault will be overlooked.

CHAPTER 10

Design of Pneumatic Systems

Design engineers involved with 'pneumatic systems' must have a basic knowledge of forces and dynamics. When moving a load by means of a cylinder – be it horizontal, vertical or through a complex angle – the engineer must have sufficient knowledge to resolve the associated static and dynamic forces. This knowledge will eventually lead the design engineer into a decision relating to cylinder dimensions, type, strut strength of the cylinder and the pressure necessary to overcome the opposing force.

10.1 Design criteria

In the design process certain criteria have to be considered no matter what type of design is being undertaken. The main design points are:

- Simplicity
- Reliability
- Efficiency
- Cost effectiveness
- Maintainability

All the above criteria are interdependent, although simplicity of design is possibly the most important. A simple design will have the least number of components; the fewer the components the less there are to fail, and so the higher the reliability.

There will be a pressure drop across each component in the system, so the lower the number of components the less the total pressure drop and associated energy losses and, consequently, the greater the system efficiency. The simpler the system, the lower the total component costs and so the better the cost effectiveness. Ease of maintenance depends on the selection of components, the layout of the system and pipework and the inclusion of test points for pressure gauges.

The designer must have a thorough working knowledge of pneumatics and full details and characteristics of all the pneumatic components available. The designer must also keep up to date with all the latest developments.

10.1.1 Design information required

The precise function of the system and its relation to and reaction with other systems in the process must be known. The designer must obtain full details or make decisions on the following points if appropriate to the system.

Actuators

Cylinders
The thrust, speed and stroke requirements for both the extend and retract stroke of each cylinder. Cylinder type (double or single acting), cylinder construction (tie-rod, non-tie-rod, cushioned or non-cushioned), cylinder mounting (foot, flange, etc.), type of air used (lubricated or non-lubricated).

Semi-rotary actuators
The torque, speed and angle of rotation requirements for both clockwise and anti-clockwise rotation. Type of construction (vane, rack and pinion, slotted screw). Type of mounting. Type of air to be used.

Motors
Static and dynamic torque requirements, speed profile for both directions of rotation. Type of motor mounting and type of air to be used.

Sequence of events
Is the sequence event or time based, or a mixture of both?

Method of control
The control system may be manual, mechanical, pneumatic, electrical, electronic or a mixture. The control may be digital or analogue.

Operating conditions
The location of the system, the environmental conditions, dirt, temperature, humidity and limitations to noise levels on any machine will all have an effect on the design considerations together with the level of operator skills and maintenance facilities.

Operating pressure
This will be limited by the available air supply. Although the compressors may deliver air at a pressure of up to 7 bar gauge there will be considerable pressure drops along the distribution network and the available pressure at the take-off points may have fallen to below 6 bar. It is usually safer to assume a working pressure of 4.5 bar and calculate actuator sizes at this pressure. It is advisable to check on site that there is an adequate flow of air at the design pressure to operate the machine satisfactorily.

Insufficient pressure may cause the machine to stall while insufficient flow will cause the machine to cycle at a reduced rate.

Special requirements
Does the exhaust air need piping to outside the work area? (This may be a requirement in some applications in food factories, etc.) Guards may need electrical or mechanical interlocks; limit switches should be fitted in such a way that they cannot be overridden. Small pneumatic cylinders can be used to operate shot pins, physically locking guards in position.

10.1.2 System to be designed

The designer must be fully conversant with the function of the system to be designed – its location, environmental conditions and the skill level of the system operators.

It is advisable for the designer to personally obtain full details of system requirements such as cylinder thrusts and speeds, motor torques, operating sequence, safety interlocks and overrides. All these considerations must be fully discussed and agreed with the client and set down as a design specification. It is essential to have a full written specification on which to base the design and to show any alterations in the customer's requirements. Care in obtaining the correct information at the start of the design process can eliminate any costly misunderstanding at the commissioning stage. Any changes in the system specification must be immediately notified to the designer.

When designing a system it is advisable to over-design the actuators to give some allowances for increases in force or torque requirements and to allow for a drop in the air supply pressure. If a compressed air distribution system is being designed, make allowances for future expansion and increases in air demand. It is more cost effective to install an oversize pipe initially than to upgrade the system at a later date.

Safety of operation is of prime importance and any failure in the system must result in the system failing safe. It must be remembered that air is a compressible media and that when air under pressure is suddenly released it expands rapidly and can be extremely dangerous. Pipelines and cylinders can store compressed air even when the supply is switched off. This stored air may be sufficient to cause an actuator to move.

When the operation of part of a circuit, or of a particular actuator, is essential to the safe operation of the system, it may be necessary to fit an in-line air receiver to store sufficient air to ensure safe operation in the case of a failure of the main air supply. When an in-line air receiver is fitted to a system it may be advisable to fit a solenoid-operated isolating valve to shut the receiver off in the case of electric failure or in an emergency situation. Each case must be carefully considered to give safe operating conditions.

An electrical pressure switch can be used to signal a loss of, or a reduction in, the air supply pressure. This signal can be used to alert the operator, stop the sequence, initiate a safety procedure, and so on. Emergency stop switches must be located at strategic positions on the machine; the exact function of the emergency stop will depend on the particular machine and process.

All moving parts, trapping points, etc., must be adequately guarded in accordance with the relevant safety regulations. Guards may be electrically interlocked or may have pneumatically powered shot pins locking the guards. Emergency stop switches may have to be fitted and their precise function determined. The emergency stop switch should be able to

(a) stop the system when the sequence is complete
(b) isolate the air supply
(c) cause the actuators to all move to a safe position.

10.1.3 Component selection and spares requirements

In determining the make of components to be used in a system, the following must all be taken into account: the importance of the system, the hours per week it is used and the make of components used in other systems in the plant. If the unit is a 'key' system it may be necessary to have a complete spare system or, alternatively, to have a complete spare set of valves and actuators.

The level of spares held in stock will depend upon the number of pneumatic systems in the plant, the level of maintenance used, the operational environment, the operators' skill and the experience gained in the operation of the machines. Should there be a well-stocked pneumatic supplier in the immediate area it may not be necessary to carry spares.

10.2 Formulae used in calculations

10.2.1 Quantity of air flow

This is measured in litres per second of free air delivered (f.a.d) throughout these calculations. Other units include cubic metres per minute and cubic feet per minute f.a.d., where

$$1 \text{ m}^3/\text{min f.a.d.} = 16.7 \text{ l/s f.a.d.}$$

$$= 35.3 \text{ ft}^3/\text{min f.a.d.}$$

Example 10.1

A pneumatic cylinder with a bore of 100 mm and a stroke of 200 mm completes 40 cycles per minute. Neglecting the piston rod volume, determine the cylinder air consumption if it is supplied at 6.5 bar.

Solution

Let, the swept volume of the cylinder per stroke be V, then

$$V = \text{Extend swept volume} + \text{Retract swept volume}$$

When the effect of the piston rod is neglected, the extend and retract swept volumes are the same. Thus,

$$V = 2 \times \text{Area of cylinder bore} \times \text{Stroke}$$

$$= 2 \times \pi \times \frac{0.1^2}{4} \times 0.2 \text{ m}^3$$

Note: All dimensions have been expressed in metres. So,

$$V = 0.00314 \text{ m}^3 \text{ per stroke}$$

The quantity of compressed air used per minute will be the swept volume per stroke times the number of strokes per minute.

$$\text{Flow rate per minute } Q = 0.00314 \times 40$$

$$= 0.1256 \text{ m}^3/\text{min}$$

To express the flow rate in items of free air it has to be multiplied by the compression ratio; i.e. the ratio of the absolute supply pressure to atmospheric pressure. The supply pressure is 6.5 bar gauge, which is 6.5 + 1, i.e. 7.5 bar absolute in this case, so the compression ratio is

$$7.5 : 1$$

Thus

$$Q = 0.1256 \times 7.5 \text{ m}^3/\text{min f.a.d.}$$

$$= 0.942 \text{ m}^3/\text{min f.a.d.}$$

$$= 15.7 \text{ l/s f.a.d.}$$

10.2.2 Flow through valves

The quantity of air flowing through any valve depends upon the pressure drop across the valve, the inlet pressure and the physical configuration of the valve. Flow control valves have an adjustable orifice to vary the flow rate through the valve. The quantity of air flowing will depend upon the upstream and downstream pressures. If a flow control valve is used to adjust the speed of a pneumatic actuator, the quantity of air flowing will depend upon the load on the actuator which governs the pressure drop over the valve. It is, therefore, impossible to obtain precise speed control under variable load conditions on a pneumatic actuator by using pneumatic flow control valves.

The quantity of air flowing through a direction control valve may be expressed as

$$Q = 6.844 C_v \sqrt{[\Delta P \times ((P_s + 1) - \Delta P)]}$$

where Q = quantity flowing in l/s (f.a.d.)

ΔP = pressure drop across the valve (bar)

P_s = inlet (supply) pressure (bar (gauge))

C_v = valve coefficient, which depends upon the valve design and its size.

The actual C_v value can be obtained from the manufacturer's catalogue. A list of values for C_v which will be used in calculations in this book is given in Table 10.1.

TABLE 10.1 Typical C_v values for a range of direction control valves.

Nominal valve size BSP	$\frac{1}{8}''$	$\frac{1}{4}''$	$\frac{3}{8}''$	$\frac{1}{2}''$	$\frac{3}{4}''$	$1''$	$1\frac{1}{4}''$
Valve coefficient (C_v)	0.25	1.0	2.0	4.0	8.0	16.0	22.0

Example 10.2

Estimate the size of direction control valve needed to supply the cylinder given in Example 10.1. The pressure drop across the valve is not to exceed 0.25 bar.

Solution

Using the formula

$$Q = 6.844C_v\sqrt{[\Delta P \times ((P_s + 1) - \Delta P)]}$$

where $Q = 15.7$ l/s f.a.d.

$\Delta P = 0.25$ bar

$P_s = 6.5$ bar

$$Q = 15.7 = 6.844C_v \times \sqrt{[0.25\ (6.5 + 1 - 0.25)]}$$

Therefore,

$$C_v = \frac{15.7}{6.844 \times \sqrt{[0.25 \times (7.5 - 0.25)]}}$$

$$= 1.7$$

A $\frac{3}{8}$ BSP valve has a C_v value of 2.0 and will have a pressure drop across it of less than 0.25 bar when driving the cylinder.

Pressure-regulating valves should be sized according to the manufacturer's information, which normally states a pressure and flow range for a particular size and design of valve.

10.2.3 Flow through pipes

Compressed air mains

It is essential that the air main linking the compressor to the point of usage is of sufficient size, otherwise there will be too high a pressure drop and air velocity in the pipe. If the pressure drop is excessive, the compressor will have to operate at a higher pressure which will require an increased power input. If the air velocity is excessive moisture separation becomes difficult, condensed moisture being driven by the air flow as a stream of water in the bottom of the pipe. A flow velocity of 5 to 8 m/s and a pressure drop not greater than 0.5 bar is reasonable in the air mains.

Example 10.3

Determine the bore of an air main to carry 6 m³/min f.a.d. at a working pressure of 6.5 bar.

Solution

Convert quantity flow rate to l/s (dm³/s):

$$\frac{6 \times 1000}{60} = 100 \text{ l/s (dm}^3\text{/s)}$$

Determine the volume of air flowing when it is compressed to 6.5 bar. Taking atmospheric pressure as 1 bar absolute, the compression ratio is $(6.5 + 1):1 = 7.5:1$. Thus, 6 m³ of free air when compressed to 6.5 bar is

$$\frac{6}{7.5} = 0.8 \text{ m}^3$$

Taking the maximum flow velocity through the pipe as 5 m/s, the pipe bore can be found as follows:

$$\text{Quantity flowing} = \text{Area} \times \text{Velocity}$$

$$\frac{0.8}{60} = \frac{\pi d^2}{4} \times 5$$

(*Note*: Units used are metres and seconds)
   Therefore,

$$d = \sqrt{\left(\frac{4 \times 0.8}{\pi \times 60 \times 5}\right)}$$

$$= 0.0583$$

$$= 58.3 \text{ mm bore}$$

The next size up standard pipe bore is 65 mm in metric sizes and $2\frac{1}{2}''$ bore imperial. If the flow velocity was increased to 7 m/s, a 50 mm bore pipe would be sufficient. If the pipe main is only of a short length, a 50 mm bore may be satisfactory.
   The pressure drop through the pipe main will probably be the deciding factor. The pressure drop through a pipe can be estimated by using the empirical formula:

$$\text{Pressure drop (bar)} = \frac{800LQ^2}{CR \times d^{5.3}}$$

where L = length of the pipe (m)

Q = volume of free air flowing through the pipe (l/s)

CR = compression ratio at the pipe entry

d = pipe bore (mm)

Example 10.4

The pipe used in Example 10.3 has a length of 300 m.
Determine the pressure drop for a 65 mm bore and a 50 mm bore pipe when $d = 65$. The flow is 100 l/s.

Solution

The compression ratio is

$$CR = 6.5 + 1 : 1 = 7.5 : 1$$

$$\text{Pressure drop 65 mm} = \frac{800 \times 300 \times 100^2}{7.5 \times 65^{5.3}}$$

$$= 0.079 \text{ bar}$$

$$\text{Pressure drop 50 mm} = \frac{800 \times 300 \times 100^2}{7.5 \times 50^{5.3}}$$

$$= 0.32 \text{ bar}$$

In this case the pressure drop of 0.32 bar would not be excessive; however, if there is any possibility of an increased demand on the air supply, it would be advisable to use the 65 mm bore pipe.

10.2.4 Pressure drop through pipe fittings

There will always be pressure drops through pipe fittings and these are usually expressed as an equivalent length of pipe of the same bore. Some typical values for both standard metric and imperial pipe bore sizes are given in Tables 10.2(a) and 10.2(b).

TABLE 10.2(a) Flow resistances of metric fittings (equivalent length in metres).

Type of fitting	Nominal pipe bore (mm)								
	15	20	25	30	40	50	65	80	100
Sharp 90° elbow	0.25	0.35	0.50	0.70	0.80	1.1	1.4	1.8	2.5
Swept 90° elbow	0.15	0.20	0.25	0.40	0.50	0.60	0.75	0.9	1.25
'Tee' straight through	0.10	0.20	0.25	0.40	0.45	0.55	0.65	0.80	1.20
'Tee' side outlet	0.50	0.70	0.90	1.40	1.60	2.10	2.70	3.70	4.90
Gate valve	0.10	0.15	0.20	0.30	0.40	0.50	0.60	0.70	0.90

TABLE 10.2(b) Flow resistances of imperial fittings (equivalent length in feet).

Type of fitting	Nominal pipe bore (inches)								
	$\frac{1}{2}$	$\frac{3}{4}$	1	$1\frac{1}{4}$	$1\frac{1}{2}$	2	$2\frac{1}{2}$	3	4
Sharp 90° elbow	1.0	1.3	1.6	2.2	2.5	3.5	4.5	6.0	8.0
Swept 90° elbow	0.5	0.6	0.8	1.1	1.3	1.7	2.3	3.0	4.0
'Tee' straight through	0.5	0.6	0.8	1.0	1.3	1.7	2.3	3.0	4.0
'Tee' side outlet	1.7	2.2	3.0	4.5	6.0	7.5	9.6	12.0	16.0
Gate valve	0.5	0.6	0.8	1.0	1.1	1.3	1.7	2.0	2.5

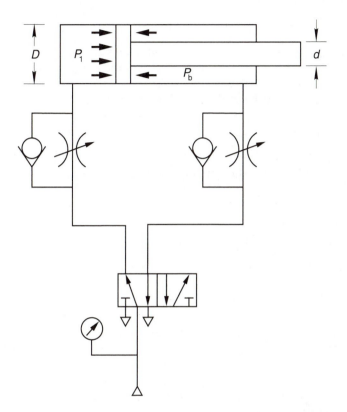

FIGURE 10.1 Arrangement of a pneumatic cylinder under dynamic conditions.

The pressure drop through the fittings in a pneumatic circuit can usually be small enough to be neglected.

10.2.5 Cylinder sizing

The static thrust developed by a pneumatic cylinder is given by:

$$\text{Static thrust} = \text{Pressure} \times \text{Piston area}$$

When the piston is moving a back pressure will be set up by the downstream flow resistance, as shown in Fig. 10.1. The pressure on the full bore side of the piston P_1 will be less than the supply pressure due to flow losses in the valves, pipes and fittings.

$$\text{Dynamic thrust} = \frac{\pi D^2 \times P_1}{4} - \frac{\pi (D^2 - d^2) \times P_b}{4}$$

It is extremely difficult to determine accurately the values of P_1 and P_b (back pressure) as they depend upon piston speed and load. It is more usual to assume that for a pneumatic cylinder:

$$\text{Dynamic thrust} = \text{Constant} \times \text{Static thrust}$$

The value assigned to the constant will depend upon the speed at which the piston is required to move. If a value of 1 is chosen, the piston movement will be very slow. It is usual to select a value for the constant of between 0.4 and 0.7, dependent upon the piston speed required; the lower the value the higher the speed.

10.2.6 Cylinder air consumption

The quantity of air used by a cylinder per stroke is the swept volume of the cylinder at the supply pressure. It is usual to state the air consumption in terms of free air – that is, at atmospheric pressure (see Example 10.1). When there is a long run of pipe between the cylinder and the control valve, the total volume of both pipes between valve and cylinder must be taken into account when calculating the air consumption of the circuit.

Example 10.5

Design study

A pneumatically operated feed unit for a carton erection machine is shown diagrammatically in Fig. 10.2. A pre-formed flat cardboard sheet is lifted from a stack of sheets by a vacuum cup attached to a pneumatic cylinder A. When cylinder A is fully retracted, cylinder B extends, transferring the sheet to a position above the carton erection machine. Cylinder A extends, pushing the sheet fully into the machine where a hot melt glue is applied and the carton erected. To facilitate this operation, cylinder A must remain fully extended for 1.5 seconds. The vacuum cups are now exhausted and the cylinder retracts, leaving the erected carton behind. The side of the erection machine is opened by extending cylinder C; the carton is then ejected onto the discharge conveyor by extending the eject cylinder D. The dimensions are:

Cylinder	Bore	Stroke
A	63 mm	400 mm
B	63 mm	800 mm
C	50 mm	300 mm
D	50 mm	600 mm

The vacuum pads are energised by switching on a supply of compressed air. The carton erection machine requires an electrical signal to start its cycle, which is completed in 1.5 seconds.

Design a suitable pneumatic system, size valves and estimate the air consumption of the unit when it is producing five cartons per minute. The air supply pressure is 6 bar gauge; assume that all cylinders operate at this pressure. The unit must be fully automatic in operation, stopping if the stack of cardboard sheets is empty or if the discharge conveyor has no space to accept a carton. The machine will automatically restart when these faults have been corrected. The cylinder control valves are to be 5/2 double-solenoid operated. A 3/2 valve, double-solenoid operated, is to be used to switch the vacuum on and off.

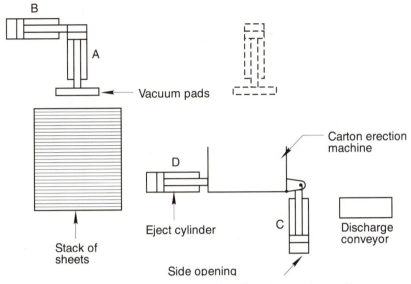

B

A

Vacuum pads

Carton erection machine

D

Eject cylinder

C

Discharge conveyor

Stack of sheets

Side opening

FIGURE 10.2 Diagrammatic representation of a carton erection machine.

Solution

Establish the essential sequence of operation of the machine. Using the notation '+' for cylinder extension and '−' for cylinder retraction, let V+ denote vacuum pads energised and V− vacuum pads de-energised. CE m/c denotes operation of the carton erection machine. At the start of the sequence all cylinders will be fully retracted with the vacuum heads de-energised.

The essential operations are:

Step   *Comments*

A+   Cylinder B must be fully retracted and there must be a sheet in the stack. Use pressure sensing to indicate that cylinder A has extended to the top of the stack and that the vacuum pad is in place.

V+   Vacuum energised. Use sensors to prove that a cardboard sheet has been picked up. An electrical vacuum pressure switch could be used.

A−   Cylinder A retracts provided that a cardboard sheet is in position. A proximity switch on the cylinder body senses when the cylinder is fully retracted.

B+   When A is fully retracted cylinder B extends to accurately locate the cardboard sheet above the carton erection machine. Prove the position of cylinder B by using an electrical limit switch or similar.

A+   Cylinder A extends, pushing the sheet into the carton erection machine. A limit switch or proximity device is used to prove that A is fully extended.

CE   Carton erection machine energised by A+ and B+ signals. Operation complete in 1.5 seconds.

V—   The vacuum can be released as soon as A is fully extended.

A—   Cylinder A retracts after it has been fully extended for 1.5 seconds. Use an electric timer. Cylinder A can be proved to be fully retracted by a proximity switch as before.

C+   Cylinder C can start to extend as soon as cylinder A is fully retracted.

D+   Cylinder D extends when cylinder C has fully extended. These operations are proved using proximity switches mounted on the cylinder barrels.

B—   This operation can occur as soon as cylinder A is fully retracted.

D—   Cylinder D can retract as soon as it has fully extended.

C—   Cylinder C can retract as soon as cylinder D has fully retracted.

At this stage in the design it is only possible to identify a sequence that will operate the machine correctly. When the machine is in production it may be necessary to adjust the sequence to optimise production rates.

One possible sequence could be:

$$A+ \ V+ \ A- \ B+ \ A+ \quad \text{Time delay} \quad A- \ C+ \ D+ \ D- \ C-$$
$$\text{CE m/c} \qquad\qquad B-$$
$$V-$$

(*Note*: Carton erection machine (CE m/c) and V— operate immediately after A+ with the time delay and then A—. Cylinder B starts to retract at the same time as cylinder C starts to extend.)

The sequence represents a complete cycle of the machine. The next cycle can start when cylinder B is fully retracted. The sequence must be fully interlocked so that a cardboard sheet cannot be fed into the carton erection machine before both cylinders C and D are fully retracted. In the sequence suggested, cylinder C starts to retract only when cylinder D is fully retracted. The interlock could be on cylinder C retract only.

Method of control

The machine is to have a relatively high production rate which would be best achieved using solenoid-operated power valves with electrical or electronic signalling. (Pneumatic signalling and pilot-operated valves are much slower in operation than the electrical equivalents.)

The control system could be a hard-wired relay logic system, which would be ideal if no further modifications were needed. In this application a degree of flexibility is required as it may be desirable to modify the sequence of operations to optimise the production rate of the machine. Such modifications would involve costly and time-consuming rewiring if a relay logic system were used. In a case like this, where some modifications in the operation may be required, it would be preferable to use a programmable logic controller. The PLC would only need a relatively small input/output capacity as only four cylinders are involved.

Sensors

The sequence is position based and sensors are needed to indicate the position of the cylinders and the operation of the vacuum pads. A timer will be used to indicate that the carton erection machine has completed its cycle.

Location	Nomenclature	Type of sensor
Cyl. A retract	a_0	Proximity switch on cylinder body
Cyl. A extend	a_1	Proximity switch on cylinder body
Cyl. A	a_p	Electrical pressure switch to indicate cylinder position on top of stack of sheets
Cyl. B retract	b_0	Limit or proximity switch
Cyl. B extend	b_1	Limit or proximity switch
Cyl. C retract	c_0	Limit or proximity switch
Cyl. C extend	c_1	Limit or proximity switch
Cyl. D retract	d_0	Limit or proximity switch
Cyl. D extend	d_1	Limit or proximity switch
Vacuum pad	$C_v +$	Electrical vacuum pressure switch or a sensitive micro-switch operated by the sheet
	$C_v -$	Sheet released
Discharge conveyor (space available)	D_{con}	Limit switch indicating no box present
Stack	SS	Indicating presence of sheet
	P_b	Start switch

The operation of proximity switches attached to the pneumatic cylinders is fully described in Chapter 8.

PLC Program

The program is to be based on the sequence already discussed, which was:

$A + C_v +$	$A - \ B +$	$A +$	Time delay $\ A - \ C + D +$	$D - \ C -$
			$C_v - \qquad\qquad B -$	
I	II	III	IV	I
$\bar{V}\ \bar{W}$	$\bar{V}\ W$	$V\ W$	$V\ \bar{W}$	$\bar{V}\ \bar{W}$

where V and W are internal relays.

Pilot signals

The next stage in obtaining the PLC program is to determine the pilot signals for the valves and internal relays, timers, etc. The pilot is given by the group identity, the previous completed operation sensor and any interlocks that have to be incorporated. Thus,

$$A + p = \bar{V} . \bar{W} . P_b . C_0 . SS$$

$$\text{also} \quad A + p = V \cdot W$$

$$A - p = \bar{V} \cdot W$$

$$\text{also} \quad A - p = V \cdot \bar{W} \cdot \text{Time delay} \cdot C_v -$$

$$B + p = \bar{V} \cdot W \cdot a_0$$

$$B - p = V \cdot \bar{W} \cdot a_0$$

$$C + p = V \cdot \bar{W} \cdot a_0$$

$$C - p = \bar{V} \cdot \bar{W} \cdot d_0$$

$$D + p = V \cdot \bar{W} \cdot c_1 \cdot D_{con}$$

$$D - p = \bar{V} \cdot \bar{W}$$

Carton erection m/c pilot $= V \cdot \bar{W}$

(Vacuum pad pilots)

$$\text{Vac} + p = \bar{V} \cdot \bar{W} \cdot a_1$$

$$\text{Vac} - p = V \cdot \bar{W}$$

Internal relay signals will be

$$V_p = \bar{V} \cdot W \cdot b_1$$

$$\overline{V_p} = \bar{V} \cdot W \cdot d_1$$

$$V \text{ disable} = \overline{\bar{V} \cdot W \cdot d_1} = V + \bar{W} + \overline{d_1}$$

$$W_p = \bar{V} \cdot \bar{W} \cdot C_v +$$

$$\overline{W_p} = V \cdot W \cdot a_1$$

$$W \text{ disable} = \overline{V \cdot W \cdot a_1} = \bar{V} + \bar{W} + \overline{a_1}$$

The timer enable is given by

$$\text{TE} = V \cdot \bar{W} \quad (K = 1.5 \text{ s})$$

PLC addresses
The addresses are as follows:

Sensor	Address
Cyl. A$-$ a_0	X0
Cyl. A$+$ a_1	X1
Cyl. B$-$ b_0	X2
Cyl. B$+$ b_1	X3
Cyl. C$-$ c_0	X4
Cyl. C$+$ c_1	X5
Cyl. D$-$ d_0	X6
Cyl. D$+$ d_1	X7
Cyl. pressure switch A_p	X10

Sheet lifted $C_v +$ X11 *Note*: the inverse of $C_v +$ may
                                               be used instead of $C_v -$, i.e.
                                               $$\overline{C_v +} = C_v -$$

Sheet released C_v X12
Sheet on stack SS   X13
Space on discharge
conveyor D_{con} X14
Stop/run button P_b X15

Internal relays addresses
V   M100
W   M101
Timer   T50

Output signal	*Address*
A+ Sol.	Y30
A− Sol.	Y31
B+ Sol.	Y32
B− Sol.	Y33
C+ Sol.	Y34
C− Sol.	Y35
D+ Sol.	Y36
D− Sol.	Y37
C_v Sol.	Y40
C_v Sol.	Y41
Carton erect m/c signal	Y42

Step ladder diagram
Using the pilot signals already obtained, a step ladder diagram can now be constructed (Fig. 10.3).

The PLC loading instructions would be as Table 10.3.

Valve sizing
The unit is required to erect five cartons per minute, giving a cycle time of 12 s. The carton erection machine requires 1.5 s for its operation, assuming it will take 0.5 s for the vacuum pads to be initiated and released, the cylinders have 10 s to operate. As electronic/electrical signalling and control is being used, the time taken for this will be negligible.

Assume that the cylinders all operate at the same linear speed on both extend and retract strokes, then the total cylinder rod travel will be:

Cylinder A: Two strokes travel 4×400 mm = 1.6 m
Cylinder B: One stroke travel 2×800 mm = 1.6 m
Cylinder C: One stroke travel 2×300 mm = 0.6 m
Cylinder D: One stroke travel 2×600 mm = 1.2 m
                               Total travel = 5.0 m

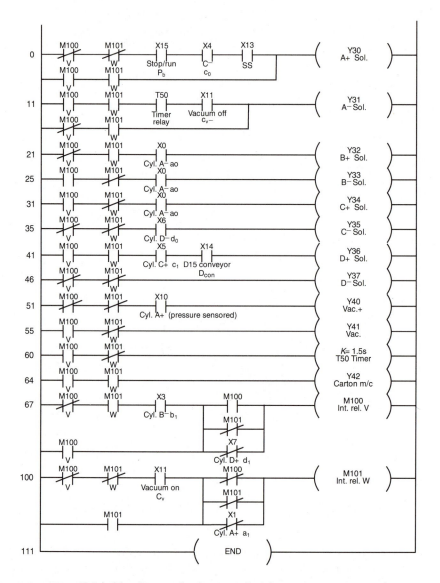

FIGURE 10.3 PLC ladder diagram for the control of the carton erection machine.

If the cylinders operate in sequence, the average cylinder piston speed will be 5.0 m in 10 s or 30 m/min. Although this is quite a high speed it will be reduced by operating cylinder B retracting at the same time as cylinders C and D extend. The extend speed of cylinder A on the stack of sheets will be regulated and cushions incorporated into the cylinders to reduce shock loads at the end of the strokes. Cylinders A and B have the largest bores, so calculate the air flow rates required by these cylinders. Maximum flow rate required will be to extend cylinder A or B, and is given by:

TABLE 10.3 PLC loading instructions for the pneumatic feed of the carton erection machine.

Step	Command and address	Comment	Step	Command and address	Comment
0	LD1 M100	Int. rel.	45	OUT Y36	D+ Sol.
1	ANI M101	Int. rel.	46	LDI M100	Int. rel.
2	AND X15	S/R button	47	ANI M101	Int. rel.
3	AND X4	C− Lt sw.	50	OUT Y37	D− Sol.
4	AND X13	Stack			
5	LD M100	Int. rel.	51	LDI M100	Int. rel.
6	AND M101	Int. rel.	52	ANI M101	Int. rel.
7	ORB		53	AND X10	A+ Press sw.
10	OUT Y30	A+ Sol.	54	OUT Y40	Vacuum+ Sol.
11	LD M100	Int. rel.	55	LD M100	Int. rel.
12	AND M101	Int. rel.	56	ANI M101	Int. rel.
13	AND T50	Time delay	57	OUT Y41	Vacuum− Sol.
14	AND X11	Vacuum off	60	LD M100	Int. rel.
15	LDI M100	Int. rel.	61	AND M101	Int. rel.
16	AND M101	Int. rel.	62	OUT T50	Timer
17	ORB		63	K1.5	
20	OUT Y31	A− Sol.	64	LD M100	Int. rel.
21	LDI M100	Int. rel.	65	ANI M101	Int. rel.
22	AND M101	Int. rel.	66	OUT Y42	Cart. err. m/c
23	AND X0	A− Lt sw.	67	LDI M100	Int. rel.
24	OUT Y32	B+ Sol.	70	AND M101	Int. rel.
25	LD M100	Int. rel.	71	AND X3	B+ L. sw.
26	ANI M101	Int. rel.	72	OR M100	Int. rel.
27	AND X0	A− Lt sw.	73	LD M100	Int. rel.
30	OUT Y33	B− Sol.	74	ORI M101	Int. rel.
31	LD M100	Int. rel.	75	ORI X7	D+ Lt sw.
32	ANI M101	Int. rel.	76	ANB	
33	AND X0	A− Lt sw.	77	OUT M100	Int. rec.
34	OUT Y34	C+ Sol.	100	LDI M100	
35	LDI M100	Int. rel.	101	ANI M101	Int. rel.
36	ANI M101	Int. rel.	102	AND X11	Vacuum CV+
37	AND X6	D− Lt sw.	103	OR M101	Int. rel.
40	OUT Y35	C− Sol.	104	LDI M100	Int. rel.
41	AND M100	Int. rel.	105	ANI M101	Int. rec.
42	AND M101	Int. rel.	106	ANI X1	A+ Lt sw.
43	AND X5	C+ Lt sw.	107	ANB	
44	AND X14	Conveyor	110	OUT M101	Int. rec.
				Clear	
			111	END	

$$Q = \pi \times \frac{0.063^2}{4} \times 0.5 \text{ m}^3/\text{s of compressed air.}$$

The supply pressure is 6 bar gauge, giving a compression ratio of:

$$(6 + 1) : 1 = 7$$

To obtain the free air flow rate multiply the compressed air required by the cylinder by the compression ratio. So the free air required is:

$$Q = \pi \times \frac{0.063^2}{4} \times 0.5 \times 7$$

$$= 0.044 \text{ m}^3/\text{s f.a.d.}$$

or

$$Q = \pi \times \frac{0.063^2}{4} \times 0.5 \times 7 \times 1000 \text{ l/s f.a.d.}$$

$$= 10.9 \text{ l/s f.a.d.}$$

To find the valve size required apply the formula for the flow of air through valves, i.e.

$$Q = 6.844 C_v \sqrt{[\Delta P(P_s + 1 - \Delta P)]}$$

Assume that the maximum allowable pressure drop through the valve is 0.25 bar, then in this case:

$$10.9 = 6.844 C_v \sqrt{[0.25(6 + 1 - 0.25)]}$$

$$= 6.844 C_v \sqrt{1.687}$$

$$C_v = \frac{10.9}{6.844 \times \sqrt{1.687}}$$

$$= 1.225$$

From Table 10.1 a $\frac{1}{4}''$ BSP nominal valve has a C_v of 1.0 and a $\frac{3}{8}''$ BSP nominal valve has a C_v of 2.0. The $\frac{3}{8}''$ valve will be suitable in this application.

In order to standardise valving manifolds and pipework use $\frac{3}{8}''$ BSP valves for all cylinders.

To control the cylinder speeds it will be necessary to fit 'in line' flow control valves or exhaust restrictors to each cylinder circuit.

The air consumption of the system will be approximately 10.9 l/s or 654 l/min. The actual consumption will be slightly less than this value as cylinders C and D have 50 mm bores and the flows were calculated for cylinders A and B which have 63 mm bores.

Note: In this solution no allowance has been made for the time delay between signalling the cylinder control valve and the piston starting to move. Nor has any allowance been made for the deceleration at the end of the cylinder strokes. The cylinder direction control valves will pass more air than is required, and so by adjusting the flow control valves in each cylinder circuit it will be possible to achieve the design target of five boxes per minute.

Pneumatic symbols

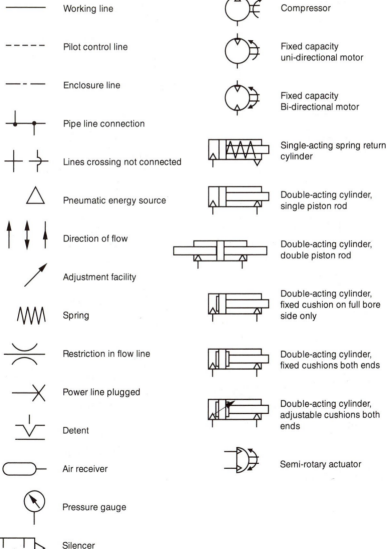

Basic components

———	Working line
- - - - -	Pilot control line
— · —	Enclosure line
	Pipe line connection
	Lines crossing not connected
△	Pneumatic energy source
↑ ↕ ↓	Direction of flow
↗	Adjustment facility
⋀⋁⋀⋁	Spring
	Restriction in flow line
	Power line plugged
	Detent
	Air receiver
	Pressure gauge
	Silencer

Motors and actuators

	Compressor
	Fixed capacity uni-directional motor
	Fixed capacity Bi-directional motor
	Single-acting spring return cylinder
	Double-acting cylinder, single piston rod
	Double-acting cylinder, double piston rod
	Double-acting cylinder, fixed cushion on full bore side only
	Double-acting cylinder, fixed cushions both ends
	Double-acting cylinder, adjustable cushions both ends
	Semi-rotary actuator

Control methods

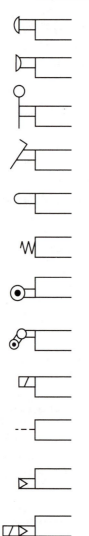

Valves

 Two port, two position

 OR Three port, two position

 Four port, two position

Five port, two position

Relief valve (adjustable)

Sequence valve (adjustable)

Pressure reducing valve (adjustable)

 Non return valve

Conditioners

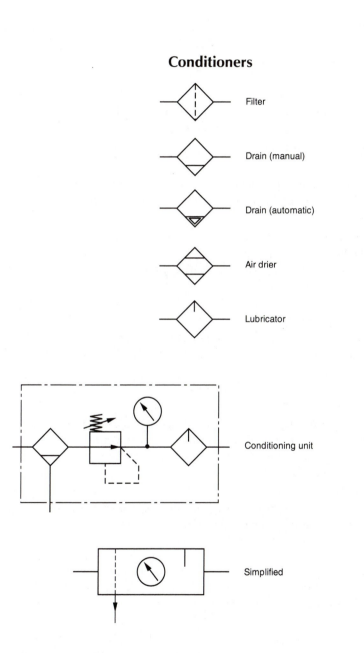

Filter

Drain (manual)

Drain (automatic)

Air drier

Lubricator

Conditioning unit

Simplified

APPENDIX 2

Questions

Standard sizes of pneumatic cylinders are to be taken from Table 4.2 (p. 115).

1 The flow demand of a system is 40 dm³/s f.a.d. at a pressure of 7 bar. Determine the nearest standard diameter of the supply pipe if the air flow velocity is not to exceed 6 m/s.

2 A pneumatic motor uses 0.5 dm³/rev of compressed air. Determine the minimum standard diameter of supply pipe when the motor is operating at 240 rev/min. The maximum air flow velocity is not to exceed 8 m/s.

3 A pneumatic system operates at a supply pressure of 6 bar, and consists of three double-acting cylinders each having a bore of 100 mm and a stroke of 200, 350 and 450 mm respectively. The total cycle time for the system is 15 s. Neglecting the effect of the piston rod and the volume of the pipework supplying the cylinders, determine the system demand. Calculate the nearest standard bore size of the main supply pipe if the air velocity is not to exceed 6 m/s.

4 A pipe conveying compressed air is 100 mm diameter and 150 m in length. Determine the flow rate through the pipe if the pressure drop is 1 bar and the entry pressure is 7 bar.
    Use the formula $P = flQ^2/d^5 P_{ave}$, where $f = 500$.
    Determine the average velocity in the pipe.

5 The exhaust air from a pneumatically operated plant in a food factory has to be piped to discharge to atmosphere outside the factory. The air consumption is 100 dm³/s f.a.d.; the discharge pipe is 50 mm diameter and 100 m long. Estimate the entry pressure at the pipe.

6 The current demand for air from a pneumatic system is 200 dm³/s f.a.d. at a minimum pressure of 4 bar. It is estimated that in the next five years demand will double to 400 dm³/s f.a.d. A compressor having a delivery of 500 dm³/s f.a.d. operating on a stop/start cycle at a maximum delivery pressure of 7 bar is to be installed.
    Determine, to the nearest m³, the size of the air receiver required so that the number of starts per hour will not exceed 20. Assume that the pressure drop in the pipework between the receiver and the plant is 0.5 bar. What will be the number of starts per hour when the system demand is half the compressor capacity?

7 A pneumatic machine requiring a supply of 200 dm³/s f.a.d. of air must complete its cycle once it has started or the product will be damaged. The cycle time is 25 seconds, the air supply pressure at the machine is 6 bar and the minimum operating pressure of the machine is 4.5 bar. A receiver is to be installed just upstream of the machine to supply it with air in case of a failure of the mains air supply. Determine the minimum size of the receiver.

8   A pneumatic cylinder with a bore of 80 mm and a rod 28 mm diameter has a stroke of 400 mm. If the cylinder completes 3 cycles/min and is supplied with air at 6.5 bar, determine the air consumption. If the air supply pressure for the return stroke is reduced to 2.5 bar, determine the saving in air consumption per minute.

9   Air is compressed from 1 bar absolute to 8 bar absolute adiabatically. If the initial air temperature is 20°C, determine the temperature after compression. Take the adiabatic index as 1.4.

10   A compressor is required to deliver 12 m³/min f.a.d. at a pressure of 8 bar. Determine the power consumption of a single-stage compressor. If a two-stage compressor was used, determine the maximum saving in power, assuming the inlet temperatures for both stages of compression are 20°C. Assume compression follows $PV^{1.3} = C$ in all cases.

11   The air inlet of a compressor is at 20°C with a humidity of 70 per cent. The compressor, which delivers 80 dm³/s f.a.d. at 7 bar gauge, has an aftercooler which reduces the air temperature to 30°C. Estimate the weight of water that has to be extracted from the compressor plant every hour.

12   A single-acting pneumatic cylinder with a stroke of 50 mm is required to clamp a component with a force of 18 kN. Determine the smallest standard cylinder to give this force when supplied with air at a maximum pressure of 7 bar. What air supply pressure is required to give exactly 18 kN and how much air will be used per cylinder stroke?

13   A double-acting pneumatic cylinder with a stroke of 500 mm is to exert a dynamic thrust of 1 kN on the extend stroke and 0.3 kN on the retract stroke. Determine the minimum size of standard pneumatic cylinder if the air supply pressure is 6 bar. Assume that the dynamic thrust is 0.6 × the static thrust. Draw a circuit to show how the cylinder is controlled and the thrust adjusted. If the cylinder has a 10 second cycle time estimate the air consumption when the cylinder thrusts are as stated.

14   A pneumatic cylinder is to be used to raise a load of 0.7 tonnes vertically through 4.0 m, the cylinder will be front flange mounted and the piston rod end pivoted and fully guided. The maximum air supply pressure at the cylinder is 6.5 bar.
    Use the formula $K = \pi^2 EJ/L^2$.
    For the buckling load K kg of a piston rod of diameter d cm, the modulus of elasticity (E) is 2.1×10^6 kg/cm². The safe working load on the piston rod is $F = K/S$, where S, the factor of safety, is 4. Determine a suitable standard cylinder assuming that the dynamic cylinder thrust is 0.6 × the static thrust.

15   A car lift is to be powered by a hydro-pneumatic cylinder. The total load that the cylinder has to lift is 1.2 tonnes through a height of 2 m. The lift must be capable of being hydraulically locked in any position. Design a suitable system using a

standard size pneumatic cylinder to raise the lift. The effective hydraulic pressure at the cylinder is 0.4 × the air supply pressure at the air–oil interface. Draw a suitable manually operated circuit and calculate the size of a suitable standard pneumatic cylinder. The air supply pressure is 7 bar maximum.

If an air compressor having a delivery of 25 l/s f.a.d. is used to supply the system, estimate the time taken for the lift to fully extend under maximum load conditions.

16   A pneumatic cylinder 80 mm bore, 25 mm rod extends against a load of 1500 N at a constant of 1 m/min under steady-state conditions. The cylinder is controlled by a 5/2 valve which has a pressure drop of 0.1 bar over each flow path. If the supply pressure is 6 bar, estimate the air flow required when using
(1) a meter-in flow control valve
(2) a meter-out flow control
as shown in Fig. A2.1
Also calculate the values of P_1 and P_2 under both conditions.

17   Two cylinders are operating in the sequence A+ B+ B− A− B+ B−. The cylinder dimensions are:

	Bore (mm)	Rod dia. (mm)	Stroke (mm)
Cyl. A	80	25	150
Cyl. B	200	40	50

The cylinders are supplied with air at 6 bar for both extend and retract strokes. If a complete cycle takes 10 seconds, estimate the quantity of free air used per minute.

18   A valve having a C_v factor of 1.7 is to be used on a system with a supply pressure of 8 bar. If the pressure drop is not to exceed 0.5 bar, determine quantity of air flowing.

19   A table having a load of 250 kg is to be raised vertically through a distance of 900 mm by a pneumatic cylinder. It is assumed that the acceleration and deceleration of the load will take place over the cushioned length (28 mm) and the load will attain a velocity of 0.8 m/s. Assume that frictional losses are equivalent to 8 per cent of the total load. The maximum pressure available is 6 bar gauge. Determine the nearest standard cylinder size and the air consumption if the cylinder operates at 10 cycles/min.

20   A pneumatic cylinder is to be used to move a mass of 5 kg. If the supply pressure is 6 bar, determine the stroke time and the maximum piston velocity, given the following catalogued data:

Cylinder $d = 50$ mm
$L = 200$ mm
$L_c = 30$ mm
Valve $C_v = 1.15$
$T_0 = 0.05$ s

Take C_e for the valve and pipework to be 2.4.

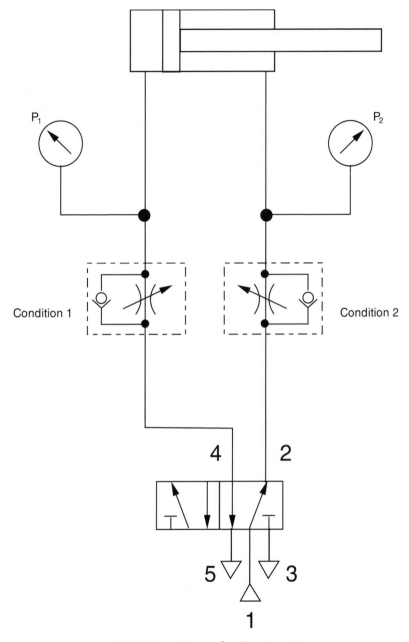

FIGURE A2.1 Diagram for Question 16.

21   The pneumatic cylinder and associated valve and pipework in the previous question is to have its load increased to 10 kg. Determine the stroke time for the new load condition.

22   A mass of 30 kg is to be raised by a vertically mounted rodless cylinder at a velocity of 1.2 m/s. The supply pressure is 10 bar and the maximum working pressure of the cylinder is 12 bar. Determine the size of cylinder that will provide adequate internal cushioning.

23   A 63 mm diameter, 6 m long rodless cylinder is to carry a load of 200 kg. Determine the minimum number of supports required and their pitch if the distance from the end of the cylinder to the centre of the carriage is 430 mm and the maximum deflection is 1.0 mm.

Answers

1. 32 mm diameter

2. 20 mm diameter

3. Demand 1.05 dm³/s
    15 mm diameter

4. 1000 litre/s

5. 1.48 bar

6. 9 m³; 19.2 starts/hour

7. 3.3 m³

8. 63.7 dm³/min f.a.d.
    25 dm³/min f.a.d.

9. 257.7°C

10. 58.2 kW; 8.9 kW

11. 2.23 kg/hour

12. 200 mm diameter
    5.73 bar
    10.6 dm³ f.a.d.

13. 63 mm diameter
    1.378 dm³/s (82.68 dm³/min)

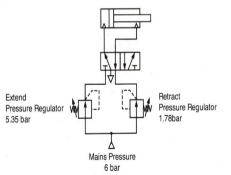

14. 160 mm bore, 40 mm rod.

15.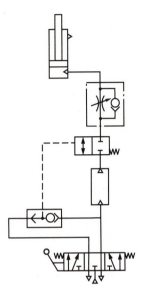

    250 mm diameter
    16 seconds

16. meter-in Q = 20.5 dm³/s
                P_i = 3.07 bar
                P_b = 0.1 bar
    meter-out Q = 34.66 dm³/s
                P_i = 5.9 bar
                P_b = 3.23 bar

17. 318.7 dm³/min f.a.d.

18. 23.27 dm³/s

19. 125 mm diameter
    24.92 dm³/s

20. 0.305 seconds
    1.41 m/s

21. 0.47 seconds

22. 50 mm diameter

23. 4 supports
    Support method ~c~ pitch = 1930 mm
    Support method ~d~ pitch = 1429 mm

Index